Bioregionalism

Bioregionalism has emerged as the new framework to study the complex relationships between human communities, government institutions and the natural world, and through which to plan and implement environmental policy. Bioregionalists believe that as members of distinct communities, human beings cannot avoid interacting with and being affected by their specific location, place and bioregion: despite modern technology, we are not insulated from nature.

Bioregionalism is the first book to explain the theoretical and practical dimensions of bioregionalism from an interdisciplinary standpoint, focusing on the place of bioregional identity within global politics. Leading contributors from a broad range of disciplines introduce bioregionalism as a framework for thinking about indigenous peoples, local knowledge, globalization, science, global environmental issues, modern society, conservation, history, education and restoration. Bioregionalism's emphasis on place and community radically changes the way we confront human and ecological issues. This book offers invaluable understanding and insights for students, activists, theorists, educators and professionals interested in ecological and bioregional topics.

Michael Vincent McGinnis is Acting Director and Postdoctoral Researcher at the Ocean and Coastal Policy Center, University of California, Santa Barbara.

Bioregionalism

Edited by Michael Vincent McGinnis

London and New York

First published 1999
by Routledge
11 New Fetter Lane, London EC4P 4EE

Simultaneously published in the USA and Canada
by Routledge
29 West 35th Street, New York, NY 10001

Typeset in Galliard by Routledge
Printed and bound in Great Britain by Redwood Books, Trowbridge,
Wiltshire

British Library Cataloguing in Publication Data
A catalogue record for this book is available from the British Library

Library of Congress Cataloguing in Publication Data
Bioregionalism / edited by Michael Vincent McGinnis,
p. cm
Includes Index.
1. Bioregionalism. 2. Environmental policy. I McGinnis,
Michael Vincent, 1962–
GE43.B56 1998   98–20210
363.7–dc21   CIP

ISBN 0–415–15444–8 (hbk)
ISBN 0–415–15445–6 (pbk)

For Christina, my strawberry girl, and the Santa
Ynez watershed

Contents

Illustrations

Tables

Figures

Contributors

Doug Aberley, M.A., is an activist, community and regional planner who teaches bioregional planning at the University of British Columbia in Canada. He is editor of *Boundaries of Home: Mapping for Local Empowerment* (1993), *Futures by Design: The Practice of Ecological Planning* (1994), and *Giving the Land a Voice: Mapping Our Home Places* (1995). He is a member of the Canadian Institute of Planners.

Tom T. Ankersen, J.D., is Staff Attorney, Center for Government Responsibility, Visiting Assistant in Law and Affiliate Faculty, Center for Latin American Studies at the University of Florida. He is coordinator of the Mesoamerican Environmental Law Program, a three-year program of applied research, training, education and advocacy in environmental law and policy.

Chet A. Bowers, Ed.D., is Professor of the Department of Curriculum and Instruction, School of Education at Portland State University in Oregon. His most recent books include: *Education, Cultural Myths and the Ecological Crisis* (1993), *Critical Essays on Education, Modernity and the Recovery of the Ecological Imperative* (1993) and *Educating for an Ecologically Sustainable Culture: Rethinking Moral Education, Creativity, Intelligence and Other Modern Orthodoxies* (1995).

David L. Feldman, Ph.D., is Senior Research Associate with the Energy, Environment and Resources Center at the University of Tennessee, Knoxville. He is author of *Global Climate Change and Public Policy* (1994), *Water Resources Management: In Search of an Environmental Ethic* (1991, 1995) and *The Energy Crisis: Unresolved Issues and Enduring Legacies* (1996), and coauthor with Curlee *et al.* of *Waste-to-Energy in the United States* (1994). He is symposium coordinator for *Policy Studies Journal*.

Dan Flores, Ph.D., is H.B. Hammond Professor of Western History at the University of Montana, Missoula. He is author of *Canyon Visions: Photographs and Pastels of the Texas Plains* (1989), *Caprock Canyonlands: Journeys into the Heart of the Southern Plains* (with E.R. Bolen; 1990), *The Mississippi Kite: Portrait of a Southern Hawk* (1993), and editor of *Jefferson and Southwestern Exploration: The Freeman and Curtis Accounts of the Red*

River Expedition of 1806 (1984) and *Journal of an Indian Trader: Anthony Glass and the Texas Trading Frontier, 1790–1810* (1985).

Bruce Evan Goldstein, M.S., is a doctoral candidate in the Department of City and Regional Planning at the University of California, Berkeley, where he is examining the use of science in regional biodiversity planning. He received his M.S. in Forest Science from the Yale School of Forestry and Environmental Studies, and has worked at the Worldwatch Institute and World Resources Institute where he assisted with the preparation of the WRI/IUCN/UNEP Global Biodiversity Strategy. He has published a number of essays on a wide range of environmental issues.

Freeman House is an author, bioregional activist and restorationist with the Mattole Restoration Council, and is author of the forthcoming *Totem Salmon* by Beacon Press.

William Jordan III, Ph.D., is publications editor at the University of Wisconsin Arboretum in Madison. He is founder and editor of the journal *Restoration and Management Notes*, and was a founding member of the Society for Ecological Restoration. He is currently spearheading the development of the New Academy for Ecological Restoration, a school without walls to train leaders for community-based ecological restoration work. He is coeditor (with Gilpin and Aber) of *Restoration Ecology: A Synthetic Approach to Ecological Research* (1987) and is completing a book entitled *The Sunflower Forest: Ecological Restoration and the New Communion with Nature.*

Daniel Kemmis is the Director of the Center for the Rocky Mountain West at the University of Montana. A former Speaker of the Montana House of Representatives and Mayor of Missoula, Montana, he is the author of *Community and the Politics of Place* (1990) and *The Good City and the Good Life* (1995).

Christopher McGrory Klyza, Ph.D., is Associate Professor of Political Science and Director of the Program in Environmental Studies at Middlebury College in Vermont. He has published a number of essays on environmental concerns, and is coeditor with Trombulak of *The Future of the Northern Forest* (1994), author of *Who Controls Public Land?* (1996), and coauthor with Trombulak of *Defining Vermont: A Natural and Cultural History* (1998). He is cofounder and board member of The Watershed Center, a group working for sustainable human and natural communities in the Lewis Creek, Little Otter Creek and New Haven River watersheds.

Ronnie D. Lipschutz, Ph.D., is Associate Professor of Politics at the University of California, Santa Cruz. He is author of *Radioactive Waste: Politics, Technology and Risk* (1980), *When Nations Clash: Raw Materials, Ideology and Foreign Policy* (1989) and *Learn of the Green World: Global Civil Society and Global Environmental Governance* (1996). He is coeditor with Ken

Conca of *The State and Social Power in Global Environmental Politics* (1993) and editor of *On Security* (1995).

Michael Vincent McGinnis, Ph.D., is Acting Director of the Ocean and Coastal Policy Center at the University of California, Santa Barbara. Since 1992, he has received grants from the US National Science Foundation to study the place of values and science in river- and watershed-based restoration. He is finishing two books entitled *On Restoring Nature, Art and the City* and *Soft Spots Between Disciplines*.

Mitchell Thomashow, Ed.D., is Director of the Antioch New England Doctoral Program in Environmental Studies. He is author of *Ecological Identity* (1995) and *KNOW NUKES: Controversy in the Classroom* (1985). He is founder and supervising editor of *Whole Terrain*.

Catherine A. Wilt, M.S., is Senior Research Associate with the Energy, Environment and Resources Center at the University of Tennessee, Knoxville. She has published a number of essays in environmental journals, and is board member of the National Recycling Coalition.

Acknowledgments

I acknowledge my hands, the dance of speech, my thoughts, my feet and my arms – that, when combined, can be the theater of a "furious rebellion." Yet these reaching thoughts often take refuge in the body. Without friends and relationships with others, each of us faces the misery of the limits of the human body and the isolated mind.

First and foremost, special thanks go to Christina for her love and endearing partnership, and Beau McGinnis for his artistic insight and alternative vision of life as poetic expression. I would also like to thank those who either read the book prospectus or chapters: John Gamman, John Clark, Hank Foster, Ed Grumbine, Andrew Light, Bryan Norton, Ari Santas, David Strong, Reed Noss, Jim Proctor and several anonymous reviewers. My editors at Routledge were also a pleasure to work with. I am grateful to Sarah Lloyd, Sarah Carty, Casey Mein and Simon Coppock for their assistance on this project. I am grateful to the National Science Foundation, which has supported my research on watershed-based restoration and management in the American West since 1992. Without this research support, the editing of this book would not have been possible.

Foreword

Certain ideas are in the air. We are all impressionable [by them], for we are made of them; all impressionable, but some more than others, and these first express them. This explains the curious contemporaneousness of inventions and discoveries. The truth is in the air, and the most impressionable brain will announce it first, but all will announce it a few minutes later.

(Ralph Waldo Emerson, "Fate")

Bioregionalism is an idea of the kind – and with the kind of potency – which Emerson describes. It is not any one person's idea, or in fact any group of people's idea. It is of a different order than that.

Earlier in the same essay, Emerson had written: "When there is something to be done, the world knows how to get it done." Bioregionalism is the world at work on itself, getting something done which the world knows to be in need of doing. It gets the work done through ideas, through words written and spoken, through organization, discipline, practice and politics. But from first to last, it is the world's work, and the world either knows or will figure out how to get it done.

It may not work as quickly as most of us would hope. Take the case of my own bioregion – the Rocky Mountain West. John Wesley Powell, a great observer of the West and later Director of the US Geological Survey argued over a century ago that the West was different, and that because of its uniqueness, it would be especially important to organize human activity in the West – including political jurisdictions – according to the lay of the land, not according to an artificial grid. We ignored Powell with a vengeance – basically trying for a century to fit the West into an undifferentiated pattern of national policies and programs, as if it were in fact no different than any place else, and then requiring it to deal with such uniqueness as may be left to it by means of political jurisdictions even less organic, less responsive to landscape than had been created in any other region.

For a century, the results of this blindness to the West's uniqueness, while damaging in a number of ways, could still be tolerated. But now, for a variety of reasons, regionalism is ripe for re-examination, and the West is positioned to begin thinking and acting in a genuinely regional manner. Watershed councils are springing up by the score across the country; bioregional efforts in places like the Greater Yellowstone or Colorado Plateau ecosystems are becoming real actors within a more and more real place called the West. It is natural to ask why, after a century of gestation, this idea which Powell expressed so passionately is finally realizing itself.

The main external contributors to this change are, on the one hand, the forces of globalization, and on the other the devolution of power downward from the national government. Globalization is favoring the emergence of organic forms at all levels: the global, the continental, the ecosystem and city-region level, and also the organic subcontinental level of regions like the Rocky Mountain West. Such regions might now begin to view devolution not simply in terms of moving authority from the national to the state level, but as an opportunity to build the capacity of organic regions to operate within the global and continental context.

All this is now upon us, or at most just around the corner. But the picture that begins to emerge is so different from what we have known that we have trouble believing it is actually happening. To gain a deeper understanding of why these ideas are now in the air (and on the ground), try viewing this picture from the perspective of the evolution of scientific theory.

Newtonian physics was the science of the nation-state. The drafters of the US Constitution were forever writing about things like billiard balls, because they were so fascinated with the strict action-and-reaction, cause-and-effect relationships they saw in the world all around them. As children of the Enlightenment, they sought to build those physical principles into the machinery of the government they were creating. Because of its high degree of predictability, which lent itself to an equally high degree of control, Newtonian physics was destined to be the science of the machine age, and of those machine-like governments which characterized that age.

As the twentieth century progressed, the radical predictability of Newtonian physics began to be assaulted by the equally radical unpredictability of quantum physics. While there remained, of course, a vast range of highly predictable phenomena, much of the universe now had to be understood as inherently unpredictable. Out of this learning grew chaos theory and then complexity theory. In a nutshell, what complexity theory tells us is that, from situations which appear to be utterly chaotic, order is constantly emerging not on our terms, but on the terms of the emerging order itself. Of all the "emergent phenomena" in the world, the most compelling is that of life itself, and the constant unfolding of life into new patterns. So, while Newtonian physics is the ideal science of the mechanical, complexity theory is the science of the organic. Which brings us back to regionalism.

Regionalism is an utterly organic phenomenon. It is never possible to tell a place that it is a region; either it is a region inherently, by its own internal logic, or it is not a region at all. As a result, regionalism stands in stark contrast and challenge to the command-and-control structures we have placed on the landscape, structures like state and county boundaries by which we attempt to tell places what they are and are not part of.

There is a further dimension of complexity theory which can help us understand this emergent phenomenon of regionalism. Fractals are those "patterns within patterns within patterns" which complexity theory has identified and which, once we become attuned to them, we see everywhere in the universe. Look on the surface of a sand dune, and you will see small sand dunes making up that surface, and even tinier dunes on the surface of the small dunes. Notice how often spirals appear in nature, from spiral nebulae to hurricanes to your draining bathtub. Complexity theory says that these forms play back and forth on one another – that what happens at a large scale is related to what happens at a much smaller scale, but not in a Newtonian way. Rather, they influence one another through that process the complexity theorists call "emergence," where new forms suddenly begin to emerge, often at several different scales at once, the way crystals emerge in a super-saturated solution.

Something very much like this could be used to describe how organic forms are suddenly emerging, just in the last decade or two, at every level of governance. Globalism is utterly organic, in the sense that the earth itself is an organism. But during the same decades that we have become so sharply aware of the organic interconnectedness of earth systems – whether ecological, economic or cybernetic – we have also begun to inhabit layers and layers of other organic forms.

Continentalism is now a fact of life. It is not going to disappear, and neither is bioregionalism, or city-regionalism, or a steadily expanding emphasis on neighborhoods. At every level, in true fractal form, we are witnessing the emergence of organic forms of human relatedness and governance.

If, as complexity theory would suggest, the earth itself has evolved into an adaptive organism which is not only the home but the living fundament of life, then it would not be altogether surprising if vast threats to life of the kind our era has produced were to cause the earth to organize itself in more life-sustaining ways. When there is something to be done, the world knows how to get it done. But it is still work, of course; it still requires careful thought, deliberation, and debate; it still requires the more or less deliberate coevolution of theory and practice. The truth is in the air, but it has to be drawn in, worked with. Something has to be made of it. This book is an exciting, stimulating, educational and most useful compilation of and reflection on this great world-work.

Daniel Kemmis

1   A rehearsal to bioregionalism

Michael Vincent McGinnis

> Home is the region of nearness within which our relationship to nature is charac-
> terized by sparing and preserving. . . . Human homecoming is a matter of
> learning to dwell intimately with that which resists our attempts to control,
> shape, manipulate and exploit.
>
> (Grange 1977: 136)

The sea anemone flows with the rhythm of the ocean's currents; its colored and
sensitive sepals feel the plentiful sea for food. Human beings are also connected
to the ebb and flow of a living earth; the sensual fibers of an animate world tug
and pull to connect culture with the land. When we listen to the landscape, we
can fall back to our primitive roots – we can smell and taste the *flesh* of the land.

In my coastal bioregion, a wealth of images, sensations and feelings are
produced by the intermingling of the sea, maritime community and landscape.
A mosaic of habitats support a number of creatures that interact to form my
community and place: bishop pine forest and tan bark oak forest; coast live oak
and riparian woodlands; chaparral–coastal sage, purple sage and coastal dune
scrub; coastal strand with freshwater or salt marsh; vernal pools and seasonal
wetlands; and blowing grasslands on coastal bluffs with rocky headlands. Shells
from nearby shores and roaming fog are part-time residents. These sights,
sounds, feelings and tastes are part of my sensual memory of place.

A healthy relationship with place is reflected in the languages spoken, the
dances and rituals of culture. Where I live, Chumash ceremonial dances such as
the Swordfish Dance, the Fox Dance, the Barracuda Dance and the Seaweed
Dance were propitiated by offerings of beads and other gifts. Each ceremonial
dance was founded upon "direct observation" of the relationship and partnership that existed between culture and nature. Each dance represented a culture's
knowledge of place. The tastes of the tribal meal were believed to be born from
the splendid place inhabited by the people. These tastes of place were celebrated
in ritual and dance. The several languages spoken by the Chumash mirrored the
ecologically and culturally diverse system. One language spoken, *Limuw*, means
"in the sea is the meaning of the language spoken." Tribal villages were named
after special places, such as *Mikiw* or "the place of mussels." The Chumash
languages exemplified a healthy maritime partnership with the landscape.

An array of human partnerships with the landscape and place are described by Paul Shepard (1996), who spent a lifetime documenting the unfolding relationships that exist between a culture, particular places and animals. "Being human," according to Shepard, "has always meant perceiving ourselves in a circle of animals" (Shepard 1996: 13). Animals have shaped human language, folklore, fairy tales, games, poetry, art, ritual, literature, myth, dancing, singing, music and religious imagery. Animals play key roles, perhaps as a totem or emblem of a family or clan, in linking humanity to place. "Inside a circle" of animals and plants, human beings are joined by a multitude of fibers that connect them to a place. Human culture is a result of this system of primordial connections with others (both plants and animals, living and nonliving).

In global economy, the cultural significance of one's place and earthly home are in jeopardy. Below the black-and-white graphic image of *pangea* (a period of earth's history when the continents were closer together) the full-page advertisement in the *New York Times* read, "Can you see the trillion dollar market?" The ad represents an economically-oriented plan, which according to a short-sighted and inebriated political and economic elite, will support a global marketplace of ideas, goods and services. The ultimate consequence of globalization on distinct places and unique cultures remains clear – we are so dramatically affecting the health of the planet that some claim we are living on a dying planet. Have you noticed the emptiness of the sky or the absence of animals missing from your place? In industrialization and large-scale economic development, ecological diversity diminishes while a culture's ability to adapt to the radical changes in ecological systems becomes increasingly doubtful. We find this same story repeated in diverse places and cultures everywhere. Planet "livability" declines for all earthly inhabitants.

This book's focus is on the place of bioregional identity within global politics. A watershed, biotic province, biome, ecosystem – in short, representations of a bioregion – can be restored and sustained if a society fosters the institutional capacity of communities to participate and cooperate to preserve the commons. Bioregionalists believe that as members of distinct communities, human beings cannot avoid interacting with and being affected by their specific location, place and bioregion: despite modern technology, we are not insulated from nature. Off the trail, the well-equipped recreational vehicle may be subdued and immobilized by "bad" weather. When hiking in bear territory, there remains a wild uncertainty.

Bioregionalism is not a new idea but can be traced to the aboriginal, primal and native inhabitants of the landscape. Long before bioregionalism entered the mainstream lexicon, indigenous peoples practiced many of its tenets (Durning 1992). Increasingly, however, population growth and new technologies, arbitrary nations/state boundaries, global economic patterns, cultural dilution and declining resources are constraining the ability of indigenous (and nonindigenous) communities to maintain traditions consistent with their past.

During the nineteenth century, the gathering of food, raising a family and the development of a community rapidly became functions of industrialized nation-states. In *Walden*, Henry David Thoreau's essay "Economy" was devoted to the importance of informal economies; those economies which support the household and community. With respect to the industrialization and formalization of economy, Thoreau writes: "Most men, even in this comparatively free country, through mere ignorance and mistake, are occupied with the factitious cares and superfluously coarse labors of life that its finer fruits cannot be plucked by them . . . he cannot afford to sustain the manliest of relations to men; his labor would be depreciated in the market. He has no time to be anything but a machine" (1995: 4). Bioregionalists remain disheartened by the ceaseless mechanization of human labor, and the general transformation of community-based economies into large-scale, formal economies which support mass production and overconsumption.

Several decades after the publication of *Walden*, another critic of industrialized society was the regional planner and theorist Lewis Mumford. In a number of books, articles and essays Mumford questioned the imperialistic and dominating character of industrial society and called for a transformation of technology and science to fit regional culture and geography (Mumford 1925; Luccarelli 1995). Mumford was one of the first proponents of ecoregionalism in the US, and he criticized the bureaucratic state as incapable of resolving the cultural and ecological crisis (Mumford 1919). Mumford combined moral and cultural criticism to empower a new ecoregionalism, which was based on an alternative phenomenology of place and a regional geography that integrated culture with nature (Luccarelli 1995: 23).

Thoreau's emphasis on the importance of the informal aspects of economy and Mumford's call for ecoregionalism are early representations of the values of contemporary bioregionalism. In a modern context based on the separation of society from the natural world, bioregionalists stress the importance of reinhabiting one's place and earthly home. A bioregion represents the intersection of vernacular culture, place-based behavior, and community. Bioregionalists believe that we should *return* to the place "there is," the landscape itself, the place we inhabit and the communal region we depend on.

This book came into being in response to the need for access to the theoretical and practical dimensions of contemporary bioregionalism – a bioregionalism that exists in a globalizing context. Societies are rapidly entering the global marketplace; individuals are participating in the telecommunicative world-wide web while the consumption and production of material goods is on the rise.

While no book can be all things to all readers, several criteria helped shape the choice of selections. Since bioregionalism is an intellectually rich and culturally diverse way of thinking and living, this compendium reflects different social and cultural aims, values (some essays are more critical of bioregionalism than others) and disciplines. Authors deal with the values of bioregionalism from a number of disciplines and epistemologies. The essays themselves manifest signs of fluctuating forces and represent diverse voices. The case studies rarely reach beyond the US. Note, however, the bioregional movement extends well beyond

North America. Each author addresses a different area of bioregional thought (linking bioregionalism as a framework for thinking about indigenous peoples, local knowledge, global environmental politics, conservation, history, education and so on). Each essay was reviewed by activists and scholars, and bears the ideological burden of the theoretical aims of the writer. Each chapter is an ideological formation with political consequences.

The first part of the book is entitled *Home place*, and represents an introduction to bioregionalism. Bioregions encompass diverse cultural areas, homelands, biodiversity, spiritual and ideological canyons, reveal economic practices, territories of the mind, unique histories of place, and geographically discrete parts of the earth. To say that you are part of a bioregion means that you inhabit a living community and place.

Since the 1960s, various publications have addressed the importance of bioregional thinking and living. This vast literature is revealed by Doug Aberley in the next chapter. Aberley's chapter is the first comprehensive history of the contemporary bioregional movement – a movement that is as deeply diverse as the landscape itself. The bioregional movement has spiritual, historical, cultural, artistic, literary and geographic identities that are very real. Bioregionalism is a grass-roots doctrine of social and community-based activism that has evolved wholly outside of mainstream government, industry and academic institutions. Bioregionalism is defined as a body of knowledge that has evolved to inform a process of transformative social change at two levels – as a conservation and sustainable strategy, and as a political movement which calls for devolution of power to ecologically and culturally defined bioregions. Aberley believes that "bioregionalism offers the best hope we have for creating an interdependent web of self-reliant, sustainable cultures."

Each bioregional move entails its own history and cultural sensibility. An understanding of the diversity of the movement is key to recognizing the likely direction that the bioregional movement may take in the context of global economy. In Chapter 3, Dan Flores argues that bioregionalism offers a range of possibilities from which a culture can make economic and lifeway choices. Flores argues that globalism fails to accept the "particularism that is the historical reality of *place*." We are not mere products of our culture or society. We are also products of the various places and contexts that we depend on. Flores emphasizes the importance of understanding one's bioregional history, which is based on a "deep time" (*longue durée*) awareness and realization of place. Flores writes: "[T]he continuing existence of [place] despite the homogenizing forces of the modern world ought to cause us to realize that one of the most insightful ways for us to think about the human past is in the form of what might be called *bioregional* histories."

In Chapter 4, I describe the values that threaten place-based and bioregional behavior. As with all earthly inhabitants, human beings are "boundary creatures"; we construct boundaries that are real and imaginary, natural and mechanical. The hybridized world of global economy is based on the development and homogenization of space (as opposed to place). The farther someone is removed from place (i.e. deplaced), the closer the bioregion resembles an environment, a natural resource, or a park. Bioregionalists believe that a viable culture must find its roots somewhere, in some place. I propose the ecological value of self-organization (or *autopoiesis*) for bioregional living.

One of the most promising moves in the direction of bioregionalism is characterized by Christopher Klyza in Chapter 5. Klyza depicts the history of watershed-based organization in Vermont. Vermont's cultures and ecosystems have survived a long period of industrialization and use. Vermont is the most rural state in the US; it has the second smallest population of any state; it has a tradition of vibrant local democracy, most apparent in town meetings; it has a strong independent streak; and its landscape is "re-wilding" and biologically recovering. Even though it is a relatively small state, it is part of three watersheds: St. Lawrence River, Connecticut River and Hudson River. Land in the state is predominantly in private ownership. Klyza describes what we can learn from the Vermont "experiments" in bioregionalism.

While Part I shows that bioregionalism originates in culture, is contingent on context and history, and on people's connections to place and the natural world, Part II (entitled *Place, region and globalism*) offers an alternative to place-based bioregional theory and practice. The authors in Part II focus on the politics of "regionalization," which is produced by political and economic identification with places. The authors focus on social networks and pluralistic identities that are emerging in globalization. Ronnie Lipschutz, Mitchell Thomashow and David Feldman/Catherine Wilt emphasize the politics of place and the region. These essays represent alternative perspectives to place-based bioregionalism.

It is important to distinguish between regionalization and bioregionalism. Transnational issues over acid rain have fostered agreements between New York and the Canadian provinces of Quebec and Ontario. Because water pollution in the river Rhine transcends nation–state boundaries, a regional agreement between France, Germany and the other European Community border states has developed. These are examples of state-sponsored regionalization, but not bioregionalism. These forms of regionalization have "no particular commitment to topographic definitions of regions" or the particularities of place (FitzSimmons 1990; Press 1995).

The form of bioregionalism described in Part I is skeptical of these forms of state-sponsored programs and initiatives. Place-based bioregionalists stress the importance of bottom-up, grass-roots and organic activities. The authors in Part II break from this form of bioregional theory and practice.

In Chapter 6, Lipschutz is *partial* to a bioregional way of describing locality, and shows that there is a prospering interregional phenomenon based on the promise of the *locale*. Contrary to some expectations, economic globalization is not paralleled by the political integration that would appear an essential condition for centralized governance. Rather, Lipschutz shows that there is emerging a parallel transnational system of rules, principles, norms and practices oriented around a very large number of often dissimilar groups and organizations. The existence of larger connections such as those created by international trade alliances and telecommunications in no way preclude the development of a series of locally-oriented networks that cut across nation-states. Places and locales are defined by watersheds, regional economies of scale, and biological life. Each place is linked in a number of social and ecological networks that Lipschutz refers to as *global civil society*. Rediscovering the importance of a river system, for example, can spawn a social network of relationships between diverse political actors who wish to protect and restore an ecosystem. Lipschutz describes the significance of *heteronomy* in the politics of global civil society – a system in which political rules are dispersed among different types of functional jurisdictions operating at local levels. Global civil society is not based on a specific place or region, but entails networks of social organizations across the world that are linked to diverse places and people.

In Chapter 7, Thomashow also offers an alternative view of place-based bioregionalism. Thomashow's argument for a "cosmopolitan" bioregionalism is consistent with Lipschutz's description of global civil society. Thomashow contends that bioregionalism speaks to the "transient" as well as the rooted, that ecological identities are broad and vast, and not necessarily linked to any specific place. Cosmopolitan bioregionalism is based on the "spirit of transregional affiliation," multiple contexts and personas, pluralistic identities, and ecological interdependencies that are found *between* places. Thomashow believes that place should be understood as a mosaic of culturally and ecologically significant signs that change, evolve and take on new meanings over time and space in a global context.

In Chapter 8, David Feldman and Catherine Wilt describe a number of regional solutions to climate change. With a focus on climate change, these authors show that states and regional governments recognize the importance of assessing impacts, formulating solutions, and identifying alternatives to reconcile natural and societal needs. Climate change policy provides a unique example of how regional solutions to global environmental problems might evolve within complicated national and international frameworks. The government-sponsored programs and initiatives that are described by Feldman and Wilt are not bioregionally-oriented. Rather, these programs and initiatives are examples of regionalization. The ultimate form of bioregional mechanisms appropriate to climate change policy, according to the authors, has yet to be determined. In short, government-sponsored programs will be required if societies are to successfully deal with the pressing problems associated with climate change.

Part III is entitled *Local knowledge and modern science*. In Chapter 9, Bruce Goldstein characterizes the relationship between modern science and local, place-based sensibilities. Many bioregionalists express reservations about scientific institutions, practices and epistemologies. Bioregionalists support a grounded, place-based knowledge. Despite the apparent tension between these two epistemologies, Goldstein argues bioregionalism should support a union of modern science and place-based knowledge. To combine scientific understanding with a place-based sensibility, Goldstein offers a strategy based on "communicative action."

The importance of a place and the region for indigenous peoples and biodiversity is, as Tom Ankersen illustrates in Chapter 10, leading to the development of several regional agreements that can unite sovereign states in Central America. Ankersen shares his experience with many of the indigenous communities who are working with international conservation organizations, scientists and governments in an attempt to preserve native lands and biological diversity. However, these plans to form reserves that link indigenous lands to areas rich in biological diversity suffer from a "conservation conundrum," which Ankersen defines as "the tension between the political and economic needs of modern society and the values of conservation, and protection of indigenous self-determination and biodiversity." The continued development of the "natural resources" that exist within these reserves goes on. The consequence of development is the significant decline in Central America's rich cultural and ecological diversity. The hope is that these reserves will eventually represent bioregions that can sustain indigenous lands and biodiversity.

In Part IV, *Toward a bioregional future*, Chet Bowers argues in Chapter 11 that a sense of place is based on an ecological literacy that combines an understanding of ecological communities (or ecosystems) and learning communities (e.g. schools). Studies show that the average child in the US can identify 1,300 corporate logos, but only ten plants and animals native to the bioregion (Lukas 1996). Bowers describes the conservative orientation of a bioregionally-oriented education and compares this orientation with modern ideology and liberal education. Liberal education is not contextually specific and seeks to nurture an array of careers and callings. Bowers argues that this instrumental view of education is a decontextualizing experience. If we are to understand the patterns of our specific bioregion, we need a new way of thinking and learning. Learning is recognized not as an end-in-itself but a process. We need to think in terms of connectedness, context and in terms of relationships (cf. Capra 1994). Bioregional education represents a shift in emphasis in liberal education and ideology to a context-driven, system-based orientation.

In the final chapter, Freeman House, William Jordan III and I describe bioregional restoration as a performative, community-based activity based on social learning and cooperation. We locate bioregionalism as a philosophy and *praxis*, and distinguish it from more scientific forms of restoration. We argue that if communities are left out of the process of restoring the landscape and place, then restoration is not bioregional. Bioregional restoration can be a therapeutic device to get us back into the "field," to foster identification with other life-forms and to *rebuild* a community with nature.

Generally, this book combines the theoretical insights of scholars and the practical wisdom of activists who offer contrasting views of the importance of place, the locale and the bioregion in global economy. The careful reader will recognize a not so subtle difference of opinions held by the authors. Thomashow and Lipschutz (and, to a degree, Flores) offer a view of bioregionalism that is linked to a constellation of places. Cosmopolitan bioregionalism seems counter-intuitive to the general value-orientations and practice of place-based bioregionalism which is described by Aberley and McGinnis. Cosmopolitan bioregionalism and global civil society requires multiple affiliations with diverse places across time and space. The chapters from McGinnis, Bowers, and McGinnis, House and Jordan propose that bioregionalism flows from the particular character of a place and context.

McGinnis and Bowers argue that modern values (liberalism, science, bureaucracy, technology) are antithetical to bioregional sensibility while Feldman/Wilt and Klyza encourage faith in conventional institutions to resolve "environmental" problems. McGinnis *et al.*, Ankersen, and Bowers propose that a tension between scientific knowledge and the more vernacular, local knowledge endemic to place and community exists while Goldstein argues that this tension between diverse epistemologies might be resolved through a more democratic, pragmatic and participatory political process.

There are other differences of opinion found in these chapters. My hope is that this book raises a number of issues and concerns that will spawn further debate and collaboration among activists.

No book can be a substitute for an acute awareness of one's place in a community, which includes the human and more-than-human world. The best introduction to bioregionalism can be found in a place-based *initiation* process that begins with one's participation in a community. Individuals are members of communities that include the direction and sound of the wind, the smell in the air, the shape of the landscape and the movement of animals. Because of this perception and sensitivity to place, human beings are able to adapt to the often subtle changes in their landscape, and find their way home. To get bioregional, humanity needs to cultivate an ecological consciousness and communal identity, and develop relationships with the neighborhood. Most neighborhoods are a mosaic of natural and mechanical elements, which may include the creek that flows to a river, a part of a mountain range or coastal zone, and a downtown city street.

Bioregional initiation requires opening up the human senses and sensibilities to the surrounding landscape; and it requires the hard work of articulating one's connection with others, "going lightly on the ground" (from Bob Dylan), meeting at the town hall, ecological literacy and self-education – all are necessary parts of the entrance into place-based service. Bioregionalism requires a long-standing adaptive orientation and cultural preference for one's place in the region and the larger region that exists beyond the horizon.

If the fact that you live and depend on a bioregion is new to you, I would prefer that you take to the wind, feel the breath of air, witness the diversity of the city or village you inhabit, and then return to this text when you are ready. Bioregionalism is the ground you walk, and it requires an acute sensibility to the uniqueness of your place. Bioregionalism is based on the fact that each place is a small world, existing for its own sake and by its own means.

As a bioregionalist, I do not believe we have lost the richness of the land, the wildest howls, the last gasp of the prairie, or the smell of animals. We have mined mountains, found a thousand wolf skulls bleached by summer shine and snow, an economy of animals, trapped. A bioregional partnership requires a new vision of life and death in the forest. Knowledge of place, within us, needs to be uncovered and revered.

References

Capra, F. (1994) *From the Parts to the Whole: Systems Thinking in Ecology and Education*, Berkeley, CA: The Center for Ecoliteracy.

Durning, A. (1992) "Guardians of the Land: Indigenous Peoples and the Health of the Earth," *Worldwatch Paper 112*, Washington, D.C.

FitzSimmons, M. (1990) "The Social and Environmental Relations of United States Agricultural Regions," in P. Lowe, T. Marsden and S. Whatmore (eds.) *Labor and Locality*, London: David Fulton Publishing.

Grange, J. (1977) "On the Way Towards Foundational Ecology," *Soundings* 60: 136.

Luccarelli, M. (1995) *Lewis Mumford and the Ecological Region: The Politics of Planning*, New York: Guilford Press.

Lukas, D. (1996) "A Place Worth Caring For," *Orion Society Notebook* Autumn/Winter: 18.

Mumford, L. (1919) "Wardom and the State," *Dial* 4: 303–5.

—— (1925) "Regions – To Live In," *Survey Geographic* 7: 91–3.

Press, D. (1995) "Environmental Regionalism and the Struggle for California," *Society and Natural Resources* 8: 289–306.

Shepard, P. (1996) *The Others: How Animals Made Us Human*, Washington, D.C.: Island Press.

Thoreau, H.D. (1995) *Walden*, Boston and New York: Houghton Mifflin, 1995. (Originally published in 1854.)

Part I
Home place

2   Interpreting bioregionalism

A story from many voices

Doug Aberley

I doubt that many people have an easy feeling about the future . . . or our ability
to protect and maintain the networks of plant and animal life upon which the
human future ultimately depends. Nor do I believe it likely that many of us
believe that the hope for the future lies in more research, or in some technological fix for the human dilemma. The research already done has produced truths
which are generally ignored. We are reaching the end of technological fixes, each
of which gives rise to new, and often more severe problems. It is time to get back
to looking at the land, water, and life on which our future depends, and the way
in which people interact with these elements.

(Dasmann 1975: 2)

Introduction

Bioregionalism is a body of thought and related practice that has evolved in
response to the challenge of reconnecting socially-just human cultures in a
sustainable manner to the region-scale ecosystems in which they are irrevocably
embedded. Over nearly twenty-five years this ambitious project of "reinhabitation" has carefully evolved far outside of the usual political or intellectual
epicenters of our so-called civilization. In urban neighborhoods, in raincoast
valleys, in prairie hollows and on semi-tropical plateaus bioregionalist communities have painstakingly and joyously learned the cultural and biophysical identity
of their home territories – their bioregions. They have also worked to share the
lessons of this hard-won experience, developing intersecting webs of bioregional
connection that now stretch across the planet. The challenging goal of this
survey is to briefly outline the remarkable history of bioregionalism.

For a number of interrelated reasons it is a difficult task to provide a definitive introduction to bioregionalism. Its practitioners protect a defiant
decentralism. There is no central committee or board of potentates that is easily
accessible for interviews or other forms of mining by journalists or academics.
The bioregional story can only be learned through long participation in local
and continental bioregion gatherings, and by assimilating ideas penned in
ephemeral journals and self-published books that rarely appear in libraries or
mass distribution outlets. It is a story best learned by listening over a very long

period of time to many voices. To complicate matters further, bioregionalism is evolving both as a body of teaching and as a social change movement at such a fast pace that it is a fool's task to identify, understand and place in proper relationship all of its dimensions.

Within the limits of my twenty years' experience as a bioregional activist and scholar I will attempt to outline the theory and practice of bioregionalism as best as possible from my own perspective. Although I have committed a considerable amount of time to thinking how to tell the story in as fair and comprehensive a manner as possible, it is ultimately only a studied opinion that I am relating. It is my hope that many others with whom I have shared the last decades of activism will tell the story from their own viewpoints. Only by allowing readers to layer what will no doubt be very different perspectives of the same events will the true layered richness of the story of bioregionalism be revealed.

This survey is restricted to review only major events and periods in the history of bioregionalism. Reference will be limited to exposition or events that have, to a degree, contributed to *expanding the borders* of bioregional thought and practice. It is important to note that these major events are not perfectly discrete, and that activists who participated in one event or episode are also active in other periods of the history of bioregionalism. It should be made absolutely clear that many layers of detail in the bioregional story have been left out. These details, which would take many hundreds of additional pages to relate, add nuance and texture to the story and are fully as important as presentation of an overall plot structure. Having given a basic orientation to the structure of this exploration, it is now possible to begin the telling of the story.

From counterculture to place-based bioregional culture

Bioregionalism gestated in the culturally turbulent decades between 1950 and the early 1970s. This era, generally labeled the "1960s," is widely perceived as a period when social, religious and political convention was confronted by a post-Second World War "Baby Boom" generation swelling through a greatly expanded post-secondary education system. Starting in the late 1940s with the North American version of the Beat Generation, a long series of interrelated social change movements were vitalized by a student-led counterculture. At the conclusion of this period there were tens of thousands of veteran social change activists in North America with experience in a variety of movements including civil rights, anti-war, peace, feminism, conservation and appropriate technology. Social historian Theodore Roszak perceptively profiled them:

> At their best, these young bohemians are the would-be utopian pioneers of the world that lies beyond intellectual rejection of the Great Society. They seek to invent a cultural base for New Left politics, to discover new types of community, new family patterns, new sexual mores, new kinds of liveli-

hood, new esthetic forms, new personal identities on the far side of power politics, the bourgeois home, and the consumer society.

(Roszak 1969: 66)

As the 1960s and the war in Vietnam wound down to their concurrent conclusion, a period of dissolution and self-reflection occurred. Three general paths of action were taken by 1960s-era activists. Individuals either (1) relinquished their activist concerns in favor of utilitarian considerations related to family, career and personal wealth generation; (2) maintained a reduced level of commitment to a succession of social change "campaigns" by the environmental movement; or (3) searched for a philosophy that described how styles of sustainable life and livelihood could be integrated with commitment to achieve a more broadly defined and fundamental degree of social and ecological change.

In tandem with the post-university diaspora of the "Baby Boom" generation, a parallel social change phenomenon was occurring in the rural regions and marginalized urban neighborhoods of North America. A new awareness evolved among residents of these communities that human and natural resources were being extracted at accelerating rates with no resulting improvement in social and environmental quality of life. As hundreds of local efforts were mounted to protest this impoverishment, often with newly located back-to-the-land and urban pioneer components of the 1960s activist community as a catalyst, a gradual new synthesis of purpose appears to have been created. A social movement was connected to the politics of home place. It is at this nexus that bioregionalism was first informally conceived, and later emerged as an important evolution in the age-old struggle to balance machine-driven economic progress with cultural and ecological sustainability.

The development of the contemporary bioregional movement includes a number of major historical events. The story of a richly diverse social and ecological movement emerged from a variety of voices which exist in a number of diverse contexts and locales. A summary of the major historical events in the contemporary bioregional movement is depicted in Table 2.1.

The complexity of events and ideas that emanate from a bioregional commitment to fundamental social change are difficult for a newcomer. The usefulness of the following broad survey is that major events in the bioregional story will be clearly revealed, and that the extensive bibliographic sources that are provided will allow access to deeper levels of exploration.

Tentative expression

The post-1960s call to create newly "indigenous activist-cultures" can be traced to the written expression of two individuals – Peter Berg and Gary Snyder. Each of these men instinctively understands that the successful growth of socially-just cultures rooted in the protection and restoration of ecosystem health requires a deep understanding of cultural tradition. The way to the future can be found by adapting genetically familiar ways of life practiced by ancestors and surviving

Table 2.1 Events in the story of bioregionalism

- Tentative expression as intersection of concern for place, politics and ecology
- Spread beyond community of origin
- Coalescence and inspiration of a vocabulary
- Attraction of an artistic, intellectual and literary vanguard
- Articulation as unified theory informed by practice
- Expression of proposed methods of applied practice
- Regional and continental congresses/gatherings
- Exploration of a broad intellectual history
- Extension of definition to more firmly include a social/spiritual dimension
- Connection/integration with other social change movements
- "Discovery" by mainstream government institutions
- Broadening into a body of teaching with balanced social and ecological dimension

---

Source: Author's own.

indigenous peoples, not in mutating humans into endlessly replaceable cogs in a machine. The focus here is on a "tribe of ecology" instead of the nation-state; a campfire circle instead of the nuclear furnace; localized rituals instead of consumerized Christmas; touch, song and shared experience instead of the narcosis of television-induced monoculture.

Snyder is best known as a Pulitzer Prize-winning poet and key participant in the San Francisco Renaissance, a West Coast manifestation of the Beat Generation. What is not as well understood is that he later became a critically important bridge between the San Francisco Renaissance and the political counterculture. Snyder's unique blending of intellectual literacy, place-centered poetics and teaching, Zen Buddhist scholarship and practice, and wilderness "savvy" are the ideal ingredients necessary for deep personal and, in many respects, cultural transformation.

Snyder's adaptation of a proto-bioregionalism first surfaces in his poetry, and in a more integrated fashion later in a widely circulated 1969 essay titled "Four Changes." After positing human overpopulation, waste and chemical pollution, and overconsumption as the root conditions of global environmental crises, Snyder pushes beyond complaint to explain how these conditions can be eliminated:

> *Goal*: nothing short of total transformation will do much good. What we envision is a planet on which the human population lives harmoniously and dynamically by employing a sophisticated and unobtrusive technology in a world environment which is "left natural." Specific points in this vision:

- A healthy and spare population of all races, much less in number than today.
- Cultural and individual pluralism, unified by a type of world tribal council. Division by natural and cultural boundaries rather than arbitrary political boundaries.
- A technology of communication, education, and quiet transportation, land-use being sensitive to the properties of each region.
- A basic cultural outlook and social organization that inhibits power and property-seeking while encouraging exploration and challenge in things like music, meditation, mathematics, mountaineering, magic, and all other ways of authentic being-in-the-world. Women totally free and equal. A new kind of family – responsible, but more festive and relaxed – is implicit.

(Snyder, in De Bell 1970: 330–1)

In a 1970 interview with Richard Grossinger in *IO* magazine, Snyder reinforces the connection he is making between place, politics and ecology as the touchstone considerations necessary to animate a new link between social activism and a sustainable life and livelihood. In explaining regionalism as a new and radically inclusive evolution in the North American social change community, Snyder believes that:

[W]e are accustomed to accepting the political boundaries of counties and states, and then national boundaries, as being some sort of regional definition; and although, in some cases, there is some validity to those lines, I think in many cases, and especially in the Far West, the lines are quite often arbitrary and serve only to confuse people's sense of natural associations and relationships. So, for the state of California . . . what was most useful originally for us was to look at the maps in the *Handbook of California Indians*, which showed the distribution of the original Indian culture groups and tribes (culture areas), and then to correlate that with other maps, some of which are in Kroeber's *Cultural and Natural Areas of Native North America* . . . and just correlate the overlap between ranges of certain types of flora, between certain types of biomes, and climatological areas, and cultural areas, and get a sense of that region, and then look at more or less physical maps and study the drainages, and get a clearer sense of what drainage terms are and correlate those also. All these are exercises toward breaking our minds out of the molds of political boundaries or any kind of habituated or received notions of regional distinctions. . . . People have to learn a sense of region, and what is possible within a region, rather than indefinitely assuming that a kind of promiscuous distribution of goods and long-range transportation is always going to be possible.

(Snyder 1980: 24–5)

Since 1970, Snyder has utilized insights gained from inhabitation of a homestead on San Juan Ridge in California's central Sierra Nevada mountains as gist for poems, interviews and essays that are distinctively bioregional in subject and texture. This expression has included poetry collections titled *Turtle Island* (1974) and *Axe Handles* (1983), interviews collected in *The Real Work: Interviews and Talks 1964–1979* (1980), and essays included in *Earth House Hold* (1969) and *The Old Ways: Six Essays* (1977). In 1990 Snyder issued an anthology of essays titled *The Practice of the Wild* that powerfully synthesized his journeyman's knowledge of syntax, his ties to a uniquely broad range of social change movements, and reflection originating from a spirited dedication to learning "home place" (Snyder 1990). Snyder's evolving versatility as a poet and essayist is accented in his most recent prose anthology, *A Place in Space* (1995). He arguably has become the single most practical proselytizer of a uniquely hybrid intellectual/spiritual/rural bioregional vision.

Peter Berg, seven years younger than Gary Snyder, arrived to live permanently in San Francisco in the early 1960s, and was active in the local experimental theater scene by 1965. After honing skills as a radical street-theater actor and playwright in the legendary San Francisco Mime Troupe he was a founding member of the legendary "Diggers," the anarcho-political conscience of the Haight-Ashbury hippie community. He became the prolific author of a series of hundreds of broadsides collectively known as the "Digger Papers," issued free in the Haight-Ashbury neighborhood between Fall 1965 and the end of 1967. One of the most celebrated of these polemics, authored by Berg, is a 23 June 1967 issue titled "Trip Without a Ticket." Berg projects an urban edginess that presages what will come later, a measured prescription for transformative change based upon connecting knowledge of place with sustained political resistance, with reintegration of human cultures into their supporting ecosystems:

> First you gotta pin down what's wrong with the West. Distrust of human nature, which means distrust of Nature. Distrust in wildness in oneself literally means distrust of Wilderness – Gary Snyder . . .
>
> Who paid for your trip?
>
> Industrialization was a battle with 19th-century ecology to win breakfast at the cost of smog and insanity. Wars against ecology are suicidal. The US standard of living is a bourgeois baby blanket for executives who scream in their sleep. No Pleistocene swamp could match the pestilential horror of modern urban sewage. No children of White Western Progress will escape the dues of peoples forced to haul their own raw materials.
>
> But the tools (that's all factories are) remain innocent and the ethics of greed aren't necessary. Computers render the principles of wage-labor obsolete by incorporating them. We are being freed from mechanistic consciousness. We could evacuate the factories, turn them over to androids, clean up their pollution. North Americans could give up self-righteousness to expand their being.

Our conflict is with job-wardens and consumer-keepers of a permissive loony-bin. Property, credit, interest, insurance, installments, profit are stupid concepts. Millions of have-nots and drop-outs in the US are living on an overflow of technologically produced fat. They aren't fighting ecology, they're responding to it. Middle-class living rooms are funeral parlors and only undertakers will stay in them. Our fight is with those who would kill us through dumb work, insane wars, dull money morality . . .
(Berg "Trip Without a Ticket," see Grogan 1990: 300–3; Halper 1991: 380; Noble 1997)

From 1967 onward Berg sustained a calculated dual commitment to act *against* machine-culture and *for* a bioregional alternative. He instigated a meta-morphosis of the Diggers into a "Free City" movement, and was instrumental in the creation and distribution of three legendary Planetedge posters that helped to irrevocably link New Left radical politics and ecological conscious-ness. Berg and dancer/actor Judy Goldhaft, who had become partners in late 1967, then moved to the Black Bear commune in the Klamath region of upper northern California, a celebrated outpost of intense social experimentation. In late 1971 the couple embarked on a journey across North America, visiting and video-taping life in a variety of counterculture communities. Their calling card – a short poem/polemic called *Homeskin* (1971) – opened with the statement: "Your body is home. Any place on this spinning geo-creature Earth is part of the skin that grows us all."

The final proto-bioregional evolution occurred in 1972 when Berg traveled to the first United Nations Conference on the Human Environment held in Stockholm, Sweden. In challenging the mainstream agenda of the conference, and in meeting and acting in concert with place-based activists from across the planet, Berg conceptualized the goals of his life's work. A common global thread of resistance and decentralized political aspiration was revealed in Stockholm – by peoples of the ethnic regions of Europe, by surviving indige-nous cultures scattered across the planet, and by emerging region-based cultures in North America.

In 1973, Berg and Goldhaft relocated and resettled in San Francisco, and worked to root the tenets of bioregionalism in the tolerant cultural medium of Bay Area counterculture society. In 1973 they founded the Planet Drum Foundation, a clearing-house for a wide variety of bioregional writing and orga-nizing activity. Between 1973 and 1979 the Planet Drum Foundation stewarded the creation of nine "Bundles" of bioregional lore. Each bundle consisted of a variety of individually printed poems, polemics, posters and essays. The first two issues of these eclectic collections were not specific to any particular locale. Later, the bundles were crafted to reflect the life and culture of specific bioregions, including the North Pacific Rim, the Rocky Mountains, and the Hudson River watershed. In 1978 the Planet Drum Foundation published an anthology of lore titled *Reinhabiting a Separate Country: A Bioregional Anthology of Northern California*. Edited by Berg, the expanded "bundle" was printed in book format with financial assistance from the California Arts Council – a granting agency of which Gary Snyder was an influential board member.

In the same period the Frisco Bay Mussel Group (FBMG), a grass-roots organization active between 1975 and 1979, became arguably the most critically important incubator for early bioregional thought and practice. The FBMG booklet *Living Here* (1977) shows how the intellectual perception of place as a focus for sustained social change activism was first related to an actual bioregional territory. Prominently featured in *Living Here* is a reverence for the ability of prehistoric human communities to adapt culture to place. This deeply rooted respect for indigenous thinking and peoples is a tenet fundamental to bioregionalism. In this period individuals including Freeman House, David Simpson, Michael Helm, Peter Coyote and a score of others debated, consented, acted and celebrated their way into a deep familiarity with bioregional thought and practice.

In 1979 the Planet Drum Foundation began publication of the biannual networking periodical *Raise The Stakes* (RTS). With a stylish layout and a stimulating mix of theoretical, practical and directory offerings, RTS remains an indispensable meeting place for a highly decentralized bioregional community of activism. In reviewing the variety and quality of organizing accomplished by Berg, Goldhaft and their many colleagues between 1967 and 1979, one notices their focused and sustained determination to introduce bioregionalism to a wider audience. This extraordinary commitment, which continues to the present, is a factor that has been critical to the success of a diverse bioregional movement.

Berg and Snyder mutually influenced each other in the period when bioregionalism was coalescing into a body of thought and teaching. Berg quoted Snyder in his early "Trip Without a Ticket" essay. Snyder was influenced by Berg when he temporarily returned to the US in 1967 from a period of isolated study in Japan. Snyder has financially supported the work of the Planet Drum Foundation through donations and as a sympathetic board member of the California Arts Council. Although living almost 200 miles apart, and having cultivated public and private personalities that reflect very different temperaments and lifestyles, Berg, Snyder and bioregionalism have coevolved in fascinating juxtaposition. The older rural communal Buddhist and the theatrical and urban radical benefit from periodic intersection. Although their lives are immensely more complicated, each activist refers to the other as validation for the commitment they both have made to promote the practice of bioregionalism.

The spread beyond community of origin

Many social change movements originated and flourished in the northern California counterculture environment, most enjoying a brief popularity that failed to extend beyond the West. In its nascent stage, bioregionalism was not

to be so confined, spreading first into the US Northwest through the exceptional writing of Freeman House and Jeremiah Gorsline. The importance of this
wider adoption of the place–politics–ecology theme cannot be overstated.
House, a friend and activist associate of both Berg and Snyder, wrote his
"Totem Salmon" essay after having relocated from San Francisco to commercially fish for salmon out of La Conner, Washington:

> Salmon is a totem animal of the North Pacific Range. Only salmon, as a
> species, informs us humans, as a species, of the vastness and unity of the
> North Pacific Ocean and its rim. The buried memories of our ancient
> human migrations, the weak abstractions of our geographies, our struggles
> towards a science of biology do nothing to inform us of the power and
> benevolence of our place. Totemism is a method of perceiving power,
> goodness, and mutuality in *locale* through the recognition of and respect
> for the vitality, spirit and interdependence of other species. In the case of
> the North Pacific Rim, no other species informs us so well as the salmon,
> whose migrations define the boundaries of the range which supports us all.
>
> (House 1974)

A year later, House and Jeremiah Gorsline, a bookseller and self-taught
forest ecologist, teamed up to add another eloquent layer to the expression of
bioregional sensibility. Their coinage of the term "future primitive" reflects a
vital extension of how the essential idea of bioregionalism is explained,
suggesting that it will be through the use of a new-old lexicon that the concept
is best passed into wider social understanding and cultural application:

> We have been awakened to the richness and complexity of the primitive
> mind which merges sanctity, food, life and death – where culture is inte-
> grated with nature at the level of the *particular ecosystem* and employs for
> its cognition a body of metaphor drawn from and structured in relation to
> that ecosystem. We have found therein a mode of thinking parallel to
> modern science but operating at the entirely different level of sensible intu-
> ition, a tradition that prepared the ground for the neolithic revolution; a
> science of the *concrete*, where nature is the model for culture because the
> mind has been nourished and weaned on nature; a *logic* that recognizes soil
> fertility, the magic of animals, the continuum of mind between species.
> Successful culture is a semi-permeable membrane between man and nature.
> We are witnessing North America's post-industrial phase right now, during
> which human society strives to remain predominant over nature. No mere
> extrapolation from present to future seems possible. We are in transition
> from one condition of symbiotic balance – the primitive – to another which
> we shall call the *future primitive*.
>
> (House and Gorsline 1974)

The spread of bioregionalism beyond the west coast of North America was assured when Gary Lawless returned to his home in Maine after spending time in California with Berg and Snyder. Lawless, a gifted poet and bookstore owner, edited an anthology of place-inspired poems, interviews, traditional songs, natural history profiles and photo essays that he self-published under the title *The Gulf of Maine: Blackberry Reader One* (1977). This work shows that bioregionalism can be transplanted from one regional place to others. Now firmly anchored on both coasts of the continent, bioregional approaches slowly began spreading inland, being adopted and adapted to meet the needs of those seeking a philosophical umbrella under which their place-centered efforts could be organized.

Coalescence and the inspiration of a vocabulary

The term bioregionalism was first conceived by Allen Van Newkirk, who had been active in eastern US radical politics, and who had met Berg in San Francisco in 1969 and again in Nova Scotia in 1971. In 1974–5, well-settled as an emigrant in Canada, Van Newkirk founded the Institute for Bioregional Research and issued a series of short papers. As conceived by Van Newkirk, bioregionalism is presented as a technical process of identifying "biogeographically interpreted culture areas ... called bioregions" (Van Newkirk 1975). Within these territories, resident human populations would "restore plant and animal diversity," "aid in the conservation and restoration of wild eco-systems," and "discover regional models for new and relatively non-arbitrary scales of human activity in relation to the biological realities of the natural landscape" (ibid.). Clear details of how these activities could be carried out were not elucidated by Van Newkirk, who, since 1975, has had virtually no influence on the idea he is responsible for naming.

The concept of bioregionalism was greatly clarified in 1977 when Berg and the renowned ecologist and California cultural historian Raymond Dasmann joined to write "Reinhabiting California," the first classic bioregional polemic. The article was originally written and published by Berg under the title "Strategies for Reinhabiting the Northern California Bioregion" (Berg 1977). Shortly thereafter, Berg was encouraged by Dasmann to submit the article for publication in the influential journal *The Ecologist*. After the piece was returned for redrafting, Berg and Dasmann worked on a major collaborative revision.

By synthesizing the experience of a cutting-edge place-based activist with that of a journeyman ecologist and experienced academic author, the bioregional vision was shown to be more than an obscure subset of the burgeoning environmental movement of the 1970s. The influence of Dasmann is obvious. At the time of his work with Berg, Dasmann was completing a seven-year United Nations-sponsored process of identifying and mapping how biophysical phenomena interact to create interlocking biogeographical territories across the planet. Dasmann was also the author of many inspirational and intellectually rigorous books, the most noteworthy being *The Destruction of California*

(1965) and *Environmental Conservation* (1984), a textbook that explored issues related to the theory and practice of "sustainability."

In merging their very different sensibilities Berg and Dasmann confidently state the enduring principles of bioregionalism by explaining the meaning of new words that bear simple, yet powerful, intent:

> *Living-in-place* means following the necessities and pleasures of life as they are uniquely presented by a particular site, and evolving ways to ensure long-term occupancy of that site. A society which practices living-in-place keeps a balance with its region of support through links between human lives, other living things, and the processes of the planet – seasons, weather, water cycles – as revealed by the place itself. It is the opposite of a society which *makes a living* through short-term destructive exploitation of land and life. Living-in-place is an age old way of existence, disrupted in some parts of the world a few millennia ago by the rise of exploitative civilization, and more generally during the past two centuries by the spread of industrial civilization. It is not, however, to be thought of as antagonistic to civilization, in the more humane sense of that word, but may be the only way in which a truly civilized existence can be maintained.
>
> *Reinhabitation* means learning to live-in-place in an area that has been disrupted and injured through past exploitation. It involves becoming native to a place through becoming aware of the particular ecological relationships that operate within and around it. It means understanding activities and evolving social behavior that will enrich the life of that place, restore its life-supporting systems, and establish an ecologically and socially sustainable pattern of existence within it. Simply stated it involves applying for membership in a biotic community and ceasing to be its exploiter.
>
> *Bioregion* refers both to a geographical terrain and a terrain of consciousness – to a place and the ideas that have developed about how to live in that place. Within a bioregion the conditions that influence life are similar and these in turn have influenced human occupancy.
>
> A bioregion can be determined initially by use of climatology, physiography, animal and plant geography, natural history and other descriptive natural sciences. *The final boundaries of a bioregion are best described by the people who have lived within it, through human recognition of the realities of living-in-place.* All life on the planet is interconnected in a few obvious ways, and in many more that remain barely explored. But there is a distinct resonance among living things and the factors which influence them that occurs specifically within each separate place on the planet. Discovering and describing that resonance is the best way to describe a bioregion.
>
> (Berg and Dasmann 1977: 399; my emphasis)

In declaring that it will be reinhabitants rather than scientists who define "home place," bioregionalism was cut forever from the tether of a more sterile biogeography. In perceiving that bioregional governance could only be established from the bottom up the bioregional movement was irrevocably put at odds with bureaucratic central government institutions (see Chapter 4). No amount of petty reform could appease a bioregional constituency that believed to the core of its collective being that democratically defined and ecologically decentralized governance was its unalienable right.

Berg and Dasmann explain how boundaries of a northern California bioregion could be defined. Their concluding judgment is that "Alta California" should be identified, both culturally and ecologically, as a separate state, a declaration that bioregionalism has an identity as a devolutionary political movement as well as that of a contemporary land ethic. Berg has utilized experience gained from intensive on-going bioregional thought and practice to write, or contribute to, important essays including "Amble Toward Continent Congress" (1976), "Devolving Beyond Global Monoculture" (1981), "More Than Saving What's Left" (1983), "Growing a Life-Place Politics" (1986), and *A Green City Program for the San Francisco Bay Area and Beyond* (Berg, Magilavy and Zuckerman 1990).

Attraction of a literary, intellectual and artistic vanguard

In always seeking new ways to express dimensions of the intent and experience of bioregionalism, key participants in other related social and cultural movements also deserve a note. Poets Gary Lawless (1977; 1994) and Jerry Martien (1982; 1984) transformed everyday experience into crystal clear lessons about how to "see" the place where you live. Social ecologist Murray Bookchin (1982) and philosophers Theodore Roszak (1975) and Morris Berman (1981) critiqued the globalist status quo and blazed trails leading to new perceptions of spiritual and cultural integration. Essayist/autobiographers Stephanie Mills (1989) and Wendell Berry (1977) used landmark events from their own lives to illustrate the challenges and opportunities to "life-in-place." The "Ecotopia" novels by Ernest Callenbach (1975; 1981) vividly portrayed how bioregion-based societies could be created and sustained. Performances by ceremonial dancers Judy Goldhaft, Alison Lang, Fraser Lang, Jane Lapiner, as well as by actor Bob Carroll, animated the unifying totemic power of water and salmon cycles in ways that no dry scientific depiction could hope to contain. These individuals, and many others, provided nascent post-1960s social change activists with a number of enticing access routes into bioregional perception and practice. Story-telling, ancient and new ritual, myth-making, theater, dance, poetry and prose all became the languages of bioregional expression.

Articulation as a unified theory informed by practice

In 1981, writer and northern California coast reinhabitant Jim Dodge synthesized a considerable body of bioregional thought, and contributed what is arguably the most compelling explanation of a bioregional vision. In a short article titled "Living By Life: Some Bioregional Theory and Practice," Dodge begins by summarizing three central values that animate bioregionalism: the importance placed on natural systems as a reference for human agency, reliance on an anarchic structure of governance based on interdependence of self-reliant and federated communities, and rediscovery of connections between the natural world and the human mind. Dodge crosses into new territory, identifying bioregionalism as more than a philosophy to live by:

> Theories, ideas, notions – they have their generative and reclamative values, and certainly a loveliness, but without the palpable intelligence of practice they remain hovering in the nether regions of nifty entertainment or degrade into flamboyant fads and diversions. . . . Practice is what puts the heart to work. If theory establishes the game, practice is the gamble.
>
> (Dodge 1981: 10)

Dodge then identifies the two broad categories of bioregional practice as being *resistance* and *renewal*. Resistance focuses against "the continuing destruction of wild systems" and "the ruthless homogeneity of national culture." Renewal is "thorough knowledge of how natural systems work, delicate perceptions of specific sites, the development of appropriate techniques, and hard physical work of the kind that puts you to bed at night" (Dodge 1981: 10–2). By adding this critical discussion of practice to what otherwise would have been yet another "New Left" or "rural populist" utopian manifesto, Dodge illuminates bioregionalism's most potent characteristic: *It is an ideal that is continuously shaped and extended through experience. It is a broad practice that begets theory, not theory stranded only in intellectual rumination and debate.*

The open and egalitarian process of defining bioregionalism, as exemplified by Dodge's writing in "Living By Life," was sustained in the pages of the previously mentioned *Raise The Stakes*, a bi-annual periodical first published by the Planet Drum Foundation in 1979. Peter Berg, Judy Goldhaft, and a revolving cast of artists, poets, writers and correspondents created a regular meeting place for the widely dispersed bioregional community. Authors and correspondents were encouraged to explain their bioregional perspective, and were empowered by the opportunity to layer their perspective and experience into the emerging mix. One of the most noteworthy issues of *Raise The Stakes*, edited by Dodge, includes submissions by seventeen contributors who offer self-criticisms relating to a variety of aspects of bioregionalism. This ability to publicly and constructively explore successes *and* weaknesses exemplifies the fact that the concept of bioregionalism is evolving through a process of place and context-driven adaptation.

Expression of methods of applied practice

The Planet Drum Foundation was instrumental in stewarding the next contemporary development in bioregional theory and practice. In a series of four short booklets written between September 1981 and January 1982 concepts geared to the practical application of a bioregional vision were articulated. *Renewable Energy and Bioregions: A New Context for Public Policy* (Berg and Tukel 1980) introduces the bioregion as a territorial container within which energy self-reliance can best be stewarded. *Reinhabiting Cities and Towns: Designing for Sustainability* (Todd and Tukel 1981) explores ecological design practice, especially as it applies to retrofitted urban centers with a variety of appropriate technology-based support systems.

In *Figures of Regulation: Guides for Re-Balancing Society with the Biosphere* (1982) Berg proposes a technique whereby "customs" can be evolved that foster evolution of lifestyles that are consciously adapted to fit the limits and opportunities of localized ecosystem processes. Taken together, these "figures of regulation" will regulate bioregion-based human societies without ideological, legal or religious coercion. The last of the small volumes, titled *Toward a Bioregional Model: Clearing Ground for Watershed Planning* (Tukel 1982), describes planning and design processes that can be used to decipher ecological carrying capacity – the parameters within which "figures of regulation" will guide cultural and economic activity in any bioregion. The meeting of these two concepts – "figures of regulation" and "bioregional model" – is expressed by Berg as:

> *Figures of regulation* is a workable phrase for the new equivalents to customs that we need to learn. Late Industrial society with its misplaced faith in technological solutions (to problems caused by unlimited applications of technology in the first place) is out of control. Our social organism is like an embryo that is suffering damage but there are no internal checks on our activities to re-establish a balance with the capacities of natural systems. The point of figures of regulation is that they would incorporate the concept that individual requirements and those of society are tied to the life processes of a bioregion. A bioregional model can identify balance points in our interactions with natural systems, and figures of regulation can operate to direct or limit activities to achieve balance.
>
> The idea of a figure as a series of movements in a dance is useful for understanding the multilayered nature of figures of regulation. The performance of a dance follows a distinct sense of rightness that would otherwise exist only as an idea, and it suggests connectedness with many other activities and ideas. It is a process that makes the invisible visible. As a dance unfolds it implies further action that is self-referenced by what has gone before. Figures of regulation are assemblages of values and ideas that can similarly become ingrained in patterns of activity.
>
> (Berg 1982: 9–10)

Regional and continental congress

A major evolution in the bioregional movement occurred in the mid-1980s, and can be attributed to the organizing skills of homesteader and appropriate technology activist David Haenke. In the late 1970s Haenke and a small group of dedicated colleagues were instrumental in establishing the Ozark Area Community Congress (OACC), the first broadly-based bioregional organization. OACC's annual congress, held every year since 1980, provided a template for the practical application of a locally-oriented and place-based bioregionalism. As word of the success of OACC spread, similar organizations were established in a growing variety of locales, first in Kansas, and later across the continent. In many cases, representatives from newly organizing bioregions would either visit OACC annual meetings, or Haenke would travel to participate in a distant inaugural gathering.

These new bioregion-based groups spawned exotically titled periodicals – *Konza* (Kansas Area Watershed Council), *Katuáh* (Bioregional Journal of the Southern Appalachians), *Talking Oak Leaves* (Seasonal Newsletter of the Ozark Area Community Congress), *Mesechabe* (Mississippi Delta Greens) and *Down Wind* (Newsletter of the Wild Onion Alliance). Each of these publications represents grass-roots bioregionalism at its best, offering a mix of local news, place-related essays, poetry, announcement of community events, and carefully thought out consideration of aspects of bioregionalism. Several memorable issues of *Mesechabe*, arguably the most eclectic bioregion-based periodical, contained a first translation of a journal made during anarchist-geographer Elisée Reclus' 1855 journey to New Orleans.

As part of his legendary role as the tireless "Johnny Appleseed" of bioregional organizers, Haenke published a booklet titled *Ecological Politics and Bioregionalism* (1984). Where earlier bioregional polemicists had been preoccupied with ecological connection actualized by a renewed anarchic primitivism, Haenke expounds a more pragmatic variant of bioregional purpose. In a tone that epitomizes mid-continent pragmatism he invokes the existence of ecological laws that will guide the positive transformation of bioregion-based societies. By adopting a style of writing that mimics the rhythm of a fundamentalist sermon Haenke describes how bioregionalism involves strict use of regenerative agriculture, appropriate technology, renewable energy sources, cooperative economics, land trusts, ecologically-based health policy, and aggressive "peace offensives." Haenke's bioregional vision is rural, practical and focused – his focus is on the politicization and institutionalization of bioregionalism.

In 1984, Haenke utilized the bioregional vision that he developed in *Ecological Politics and Bioregionalism* as a framework for organizing and convening the first North American Bioregional Congress. Over 200 participants from several continents were attracted to this landmark event, in which policies in twenty-three areas of bioregional concern were developed by committees, debated in plenary sessions, adapted as deemed necessary, and adopted by consensus. These policies are depicted in Table 2.2. The written record of this gathering, *North American Bioregional Congress Proceedings* (Henderson *et al.* 1984), as well as the proceedings of four bi-annual Continental Congresses/Gatherings that have followed (Hart *et al.* 1987; Zuckerman 1989; Dolcini *et al.* 1991; Payne 1992), are key sources that reveal how the concept of bioregionalism has expanded. A second vital source of bioregional history emanating from the continental congresses are daily newsletters issued under the name *Voice of the Turtle*. Each issue summarizes reports from the previous day's events, as well as a variety of poems, personal statements and related important contextual material.

The published proceedings of congresses and gatherings held in scores of individual bioregions provide detail regarding ways in which the definition of bioregionalism has been adapted to suit the needs and nuance of different cultural and biophysical settings. Noteworthy publications, among many others, include *Kansas Area Watershed (KAW) Council Resolutions* (Kansas Area Watershed Council 1982), *The Second Bioregional Congress of Pacific Cascadia: Proceedings, Resources and Directory* (Scott and Carpenter 1988) and *Proceedings: First Bioregional Congress of the Upper Blackland Prairie* (Marshall 1989).

Table 2.2 North American Bioregional Congress (NABC) Committee Structure (1984–90)

Committee	NABC1	NABC2	NABC3	NABC4
Agriculture/permaculture	X	X	X	X
Bioregional education	X	X	X	X
Bioregional movement	X	X	X	X
Children's	—	—	—	X
Communication/media	X	—	X	X
Communities	X	—	X	—
Culture and arts	X	X	X	X
Ecodefense	X	X	X	X
Ecofeminism	X	X	X	X
Economics	X	X	X	X
Evolving leadership	—	—	X	—
Forests	X	X	X	X
Green cities	—	X	X	X
Green movement	X	X	X	X
Health	—	X	X	X
Indigenous peoples	—	—	X	—
MAGIC (Mischief, Animism, Geomancy and Interspecies Communication)	—	X	X	X
Mapping	—	—	—	X
Materials reuse/toxic waste	—	—	X	—
Native peoples/people of color	X	X	X	X
Spirituality	X	X	X	—
Transportation	—	—	—	X
Water	X	X	X	X

Source: Author's own.

Exploration of a broad intellectual history

In 1985, the Sierra Club published *Dwellers in the Land: The Bioregional Vision*, authored by respected cultural historian and bioregionalist Kirkpatrick Sale. In presenting bioregionalism for the first time to a mass literary audience he argues that:

- Machine-based civilization has abandoned the Greek mythological concept that the earth, Gaia, is a single sentient organism.
- As a result, multiple social and ecological crises exist that threaten the survival of human civilization.
- Bioregionalism offers an alternative paradigm based upon principles including:
  - Division of the earth into nested scales of "natural regions"
  - Development of localized and self-sufficient economies
  - Adoption of a decentralized structure of governance that promotes autonomy, subsidiarity and diversity
  - Integration of urban, rural and wild environments

- Bioregionalism is connected to anarchist, utopian socialist and regional planning traditions.

Sale's treatise is instrumental in introducing bioregionalism to the general public in two fundamental ways. First, Sale greatly expands upon Dodge's presentation of bioregionalism as a unified theory, or in Sale's terminology, as a "paradigm." Table 2.3 depicts the structure of the bioregional paradigm described by Sale.

Table 2.3 Events in the story of bioregionalism

	Bioregional paradigm	*Industrio-scientific paradigm*
Scale	Region	State
	Community	Nation/world
Economy	Conservation	Exploitation
	Stability	Change/progress
	Self-sufficiency	World economy
	Cooperation	Competition
Polity	Decentralization	Centralization
	Complementarity	Hierarchy
	Diversity	Uniformity
Society	Symbiosis	Polarization
	Evolution	Growth/violence
	Division	Monoculture

Source: Sale (1985: 50)

Second, Sale shows that the values of bioregionalism existed in the works of North American and European regionalists. Citing classic sources in regional planning history, including Carl Sussman's *Planning the Fourth Migration: The Neglected Vision of the Regional Planning Association of America* (1976) and Friedmann and Weaver's *Territory and Function: The Evolution of Regional Planning* (1979), Sale identifies Frederick Jackson Turner (1861–1932), Howard Odum (1884–1954) and Lewis Mumford (1895–1990) as progenitors of American regionalism. Sale ties American regionalist thought to the earlier related European expression of Frédéric Le Play (1806–1882), Friedrich Ratzel (1844–1904), Paul Vidal de la Blache (1845–1918) and Patrick Geddes (1854–1932).

In tying bioregionalism to a 200-year tradition of resistance against machine- and metropolitan-dominated culture, Sale creates both challenge and opportunity. A challenge in that these relatively obscure intellectual and activist traditions required exploration so that the lessons of their successes and failures could be understood. An opportunity in that bioregionalism could be viewed as only the latest reincarnation of a centuries long effort to define how socially-just and ecologically sustainable human cultures could be created and sustained. Sale single-handedly attempts to characterize the intellectual genealogy of bioregionalism. The Sierra Club's book distribution network, and Sale's reputation as a respected cultural historian, ensured that *Dwellers* gained a much higher profile than any book on bioregionalism published before or since.

Dwellers became a lightning-rod for criticism from sources both within and outside the bioregional movement. Reviewers from inside the bioregional movement resented Sale's supposedly "dressed-up" intellectualization, and the lack of exposure given to less definitive and more anarchic strands of the bioregional vision (Helm 1986: 12; LaChapelle 1988: 183–4). This criticism, to some extent unjustified in light of Sale's clear statement that the book represents only his own studied opinion, reacts against even the hint that any one interpretation of bioregionalism could be presented as being definitive. It also points to the existence of some level of tension between the most active bioregional theorists and organizers. *This tension, which remains today, appears to have evolved as an impediment against any single individual becoming a movement leader or independent spokesperson.*

The issue of leadership in the highly decentralized bioregional movement bears further comment. Leadership is critically important to the success of any social change movement that confronts an opponent as insidiously powerful as globalism. Bioregionalists temper this understanding by remembering the fate of 1960s-era leaders who either succumbed to the vainglory of media-created charisma or treated dogmatic allegiance to indulgent rhetoric as more important than empowering a self-actualized citizenry. The compromise that seems to have been accepted is that leaders at the bioregional level will most likely be those who best put to practice of locally-focused resistance and cultural renewal.

Academics and ideologues attempting to tie bioregionalism into a variety of their debates have also made Sale's treatise a target for critique. A volume that serves as an initial hopeful statement of purpose and that introduces bioregion-alism in an accessible manner does not fare well when dissected by reviewers who are comfortable with the intricacies of post-Marxist, academic anarchist, planning, or other variants of often obscure post-modern discourse. An initial *Dwellers*-inspired review of bioregionalism by James Parsons, protégé of revered cultural geographer Carl Sauer, was extremely positive (Parsons 1985). Reviewers have since been less supportive of Sale's bioregional vision.

Journal or anthology articles by Alexander (1990; 1993), Atkinson (1992), McTaggart (1993) and Frenkel (1994) have all used *Dwellers*, and usually a relatively limited number of other references, to describe bioregionalism in essentially a simplistic manner. These authors are intent to squeeze, or in Atkinson's view, to "sharpen" bioregionalism so that it properly fits into the framework of their narrower disciplinary interests in planning or geography. These articles are written in a tone that communicate a hopeful interest in bioregionalism's future, but only if the concept can be perfectly purged of a variety of weaknesses, including that it is potentially or inherently reductionist, utopian, ahistoric, or ecologically deterministic.

A final form of writing in which *Dwellers* is referenced includes what can be labeled sustainability manifestos written by popular social theorists, for example Milbrath's *Envisioning a Sustainable Society: Learning Our Way Out* (1989) and Rifkin's *Biosphere Politics: A New Consciousness for a New Century* (1991). In several pages within much longer works, bioregionalism is presented primarily as proposing the concept of a useful territorial container, the bioregion. In all the books and articles in which it is mentioned, *Dwellers In The Land* remains an influential, and controversial, source of bioregional lore.

Bioregionalism is best understood when viewed from the "inside," not from reading one or several texts. Gatherings should be attended, ephemeral periodi-cals reviewed, restoration projects participated in, and place-based rituals and ceremonies shared. Examples of critical appraisals which successfully adopt this approach can be found in the pioneering graduate theses of Aberley (1985) and Carr (1990). Aberley details the historic exploitation of a rural bioregion, then explores how a bioregional alternative might be implemented. Carr interprets the social and philosophical evolution of bioregionalism based on a decade of taped interviews and wide participation in bioregional events.

Bioregionalism did not emerge in the 1970s perfectly conceived or practiced. Intense and informed debate about the weaknesses of the essential tenets of bioregional living is constant. Without recognizing the diversity of voices from which bioregionalism originates, and the context-driven manner in which the bioregional movement evolves, academic critiques can only be short-sighted and reductionist.

Extension to include social/spiritual definition

Another major development in the theory of bioregionalism is Thomas Berry's *The Dream of the Earth* (1988), a collection of essays joined by a bioregional theme. A theologian active in the Hudsonia bioregion in New York State, Berry is concerned with constructing a bioregional world-view firmly linking spirituality with a form of social organization. Berry describes a set of six "functions" which are necessary for bioregional living:

> The first function, self propagation, requires that we recognize the rights of each species to its habitat, to its migratory routes, to its place in the community. The bioregion is the domestic setting of the community just as the home is the domestic setting of the family . . .
>
> The second bioregional function, self-nourishment, requires that the members of the community sustain one another in the established patterns of the natural world for the well-being of the entire community and each of its members. Within this pattern the expansion of each species is limited by opposed lifeforms or conditions so that no one lifeform or group of lifeforms should overwhelm the others . . .
>
> The third function of a bioregion is its self-education through physical, chemical, biological, and cultural patterning. Each of these requires the others for its existence and fulfillment. The entire evolutionary process can be considered as a most remarkable feat of self-education on the part of the planet Earth and of its distinctive bioregional units . . .
>
> The fourth function of a bioregion is self-governance. An integral functional order exists within every regional life community. This order is not an extrinsic imposition, but an interior bonding of the community that enables each of its members to participate in the governance and to achieve that fullness of life expression that is proper to reach . . .
>
> The fifth function of the bioregional community is self-healing. The community carries within itself not only the nourishing energies that are needed by each member of the community; it also contains within itself the special powers of regeneration. This takes place, for example, when forests are damaged by the great storms or when periods of drought wither the fields or when locusts swarm over a region and leave it desolate. In all these instances the life community adjusts itself, reaches deeper into its recuperative powers, and brings about a healing . . .
>
> The sixth function of the bioregional community is found in its self-fulfilling activities. The community is fulfilled in each of its components: in the flowering fields, in the great oak trees, in the flight of the sparrow, in the surfacing of the whale, and in any of the other expressions of the natural world. . . . In conscious celebration of the numinous mystery of the universe expressed in the unique qualities of each regional community, the human fulfills its own special role. This is expressed in religious liturgies, in market festivals, in the solemnities of political assembly, in all manner of play, in music and dance, in all the visual and performing arts. From these come the cultural identity of the bioregion.
>
> (Berry 1988: 166–8)

Similar to Sale, Berry supplies an extensive bibliography as an integral part of his book. By availing easy access to the intellectual underpinnings of their exposition, both Sale and Berry reinforce the fact that bioregionalism is connected to a larger and much deeper philosophical tradition than its most recent counterculture incarnation might indicate.

The characterization of the spiritual importance of bioregionalism has occurred in two other important areas. In Texas, Joyce and Gene Marshall grafted a radical Christian tradition with bioregionalism to create a dynamic grass-roots activist–spiritual movement whose work is expressed in the pages of a periodical titled *Realistic Living*, first published in 1985. On a parallel path, deep ecology's earth spirituality has been adopted by bioregionalists who experiment with meditation, vision questing, celebration of seasonal cycles, and a host of other rituals. Inspirational books in this genre include *Thinking Like a Mountain: Towards a Council of All Beings* (Seed *et al.* 1988), *Sacred Land, Sacred Sex: Rapture of the Deep* (LaChapelle 1988) and *Truth or Dare: Encounters with Power, Authority, and Mystery* (Starhawk 1987).

Connection/integration with other social change movements

Since the late 1980s, the development of contemporary bioregionalism has not evolved so much in broad strokes as it has from an organic and incremental process driven by the experience of a spreading network of activists and organizations. The process of bioregional dissemination and experimentation, although difficult to trace, represents the real current strength of the diverse bioregional movement. In hundreds of towns, cities and rural enclaves, a parallel movement that supports bioregional governance is quietly and persistently taking root.

Bioregionalists need to explore their intellectual and practical relationships to a host of other vital social and ecological movements. No single movement can succeed in inspiring transformation of the "consumer–producer society" on its own. Nor can a single movement overcome the politics of displacement and isolationism endemic to globalization. The bioregional movement remains open and inclusive. Bioregionalism embraces the values expressed in ecofeminism (Muller 1984; Plant 1986), earth spirituality (LaChapelle 1988), permaculture (Crofoot 1987), ecological restoration (House 1974; 1990), among others. This integration is reflected in an essay by Michelle Summer Fike and Sarah Kerr, who write:

> Bioregionalism and ecofeminism are two streams of the contemporary environmental movement that provide related yet distinct frameworks for analyzing environmental and social justice issues, as well as offering visions of more sustainable ways of living with the earth. Seeing the linkages between feminism, environmentalism, anti-racism, gay liberation, peace and justice work, and all of the other struggles for freedom and democracy is

critical to our work as community activists and organizers. We feel that a greater understanding of these interconnections is one of the most important lessons offered by a joint examination of ecofeminism and bioregionalism.

(Fike and Kerr 1995: 22)

Evidence of this process of constant connection and integration can also be found in the previously introduced published proceedings of six North American Bioregional Congresses/Turtle Island Bioregional Gatherings held since 1984, in the pages of twenty-five bi-annual issues of the Planet Drum Foundation's networking and bioregional theory periodical *Raise The Stakes*, and in a growing variety of journals which carry articles with bioregional themes.

One future goal of bioregionalism is to successfully integrate with other social change movements (e.g. the environmental justice movement) to ensure that a more potent ability to affect social, political and ecological transformation can be achieved. Perhaps the greatest hope for bioregional activity lies in this integration with other movements. Bioregionalism supports place-based cultural transformation. The bioregion could become the political arena within which resistance against ecological and social exploitation could be produced.

Mainstream "discovery" and (mal)adaptation

In the early 1990s, bioregionalism was "discovered" by politicians, natural resource managers and environmental policy-makers who primarily serve government institutions and corporate interests. In a range of national settings, the language of bioregionalism has been appropriated to assist in conceptualizing experiments in institutional and organizational reform. However, these government-sponsored developments have occurred *with little reference to or contact with the grass-roots bioregional movement*. Explicit uses of bioregional terminology include the September 1991 *Memorandum of Understanding* signed between heads of federal and state resource management agencies active within California state borders (California State Resources Agency 1991). In Ontario a joint Provincial–Federal task force identified a "Greater Toronto Bioregion" as best enabling management of a large metropolitan area (Royal Commission on the Future of the Toronto Waterfront 1992). Each of these initiatives have defined bioregion borders from the top down, and have not adequately explained the role communities should play in these alternative territorial regimes.

Implicit adoption of bioregional tenets include the restructuring of regional governance units in New Zealand to match major watershed boundaries (Furuseth and Cocklin 1995; Wright 1990). In Nunavut, a new ethnic bioregion is to be proclaimed in the eastern Canadian Arctic in 1999; a man and a woman will be chosen to represent each new electoral area (Devine 1992). Similarly, the Navajo Nation is evolving a "dependent sovereignty" relationship within its host jurisdiction, the United States of America (Commission on Navajo Government Development 1991). In Europe, a Committee of the Regions has since 1994 provided nearly 100 traditional ethnic bioregions with a recognized policy-proposing forum (European Communities 1994). In the Great Lakes, Gulf of Maine and Cascadia, scientific and planning panels have adopted bioregions as the territorial unit within which a variety of diverse planning activities will be focused.

Ideas central to the bioregional vision have been adopted by mainstream institutions. This appropriation of bioregional values can be considered a compliment to the relative strength of the movement. At the same time, however, these initiatives are generally devoid of a crucial bioregional value – the redistribution of decision-making power to semi-autonomous territories who can adopt ecological sustainable *and* socially-just policies. Bioregionalists fear that the general public will identify bioregionalism with these rhetorical and pragmatic government-sponsored initiatives rather than associate bioregionalism with its grass-roots and organic origin.

Broadening into a body of teaching

The latest phase in the development of contemporary bioregionalism materialized in the 1990s. After theoretical expression of techniques of applied bioregionalism were issued by the Planet Drum Foundation in the early 1980s, a long period of isolated experimentation with these methods and others took place. As this experimentation matured, and as the number of individuals and organizations interested in bioregionalism increased, the need arose to provide the means by which experience with tested techniques of applied bioregionalism could be more widely explained and taught.

In 1990 New Society Publishers, centered on Gabriola Island in British Columbia, reacted to this need by initiating two important publishing projects. The first involved assembling a definitive anthology of the best representative sample of available writing on bioregional theory and practice. As conceptualized by an immensely literate team of editors including bioregional movement veterans Van Andruss, Eleanor Wright, and New Society principals Judith and Christopher Plant, *Home! A Bioregional Reader* (1990) deftly layers bioregionalism's many themes into a seamless whole. *Home!* remains the single most convenient and comprehensive way to read oneself into familiarity with the bioregional vision.

New Society Publishers' second pioneering effort involved founding of the *New Catalyst Bioregional Series*. In a format that allows a knowledgeable editor to weave together summaries of bioregional thought and practice emanating from a variety of geographical and gender perspectives, the *Bioregional Series* has become an indispensable source of cutting-edge bioregional lore. In eight editions the *Bioregional Series* has explored individual themes including interviews with key bioregional thinkers (Plant and Plant 1990), green economics (Plant and Plant 1991), community empowerment (Plant and Plant 1992), human community–ecosystem interaction (Meyer and Moosang 1992), community-based alternatives to alienation (Forsey 1993), bioregional mapping (Aberley 1993), ecological planning (Aberley 1994), and exploration of the ecological footprint method of measuring a community's appropriation of ecological capital (Wakernagel and Rees 1996). Recent books by other publishers, including *Giving the Land a Voice: Mapping Our Home Places* (Harrington 1995) and *Discovering Your Life-Place: A First Bioregional Workbook* (Berg 1995), have added to the growing range of "how to" material available to practicing bioregionalists.

Attempting a synthesis

The challenges of twenty years of continuous extension of purpose has stretched the ability of a highly decentralized movement not only to guide its own growth process, but also to communicate its principles in a timely, purposeful, and clear manner. Consequently, the tenets of bioregionalism, and the rich history of the bioregional movement, are not as widely known by the general public as are those of other contemporary social change movements. It is possible that this relative obscurity is about to change. The publication of this volume, as well as a growing number of similar books being written by experienced bioregional activists, indicates that a formative, inwardly-focused organizational period of development may be at an end.

Ironically, bioregionalism's greatest strength stems from the fact that it *has* remained relatively obscure. The goal of the bioregional theorist has been to reflect on the needs and values of living-in-place, not to craft a seamless theoretical construction or utopian diatribe. As a loosely bundled collection of ideals which emanate from the reflective experience of place, bioregionalism "speaks" to social change activists tired of convoluted ideological dogma. Bioregionalism is a daringly inclusive doctrine of fundamental social change that evolves best at the level of decentralized practice. Although none of the tenets of bioregionalism are etched in stone, these tenets stake claim to a dynamic, grass-roots approach to conceptualizing and achieving transformative social change:

Bioregional world-view

- Widespread social and ecological crises exist; without fundamental change preservation of biodiversity, including survival of the human species, is in doubt.
- The root cause of these threats is the inability of the nation-state and industrial capitalism – patriarchal, machine-based civilization rising from the scientific revolution – to measure progress in terms other than those related to monetary wealth, economic efficiency or centralized power.
- Sustainability – defined as equitably distributed achievement of social, ecological and economic quality of life – is better gained within a more decentralized structure of governance and development.

- The bioregion – a territory revealed by similarities of biophysical and cultural phenomenon – offers a scale of decentralization best able to support the achievement of cultural and ecological sustainability.

Culture

- Both humans and other species have an intrinsic right to coevolve in local, bioregional and global ecosystem association.
- Bioregion-based cultures are knowledgeable of past and present indigenous cultural foundations, and seek to incorporate the best elements of these traditions in "newly indigenous" or "future primitive" configurations.
- Bioregion-based culture is celebrated both through ceremony and vital support of spiritual reflection and related cultural arts.

Governance

- Bioregion governance is autonomous, democratic and employs culturally-sensitive participatory decision-making processes.
- Political and cultural legitimacy are measured by the degree to which a steward achieves social and ecological justice, and ecosystem-based sustainability.
- Intricate networks of federation will be woven on continental, hemispheric and global bases to ensure close association with governments, economic interests and cultural institutions in other bioregions.

Economy

- Human agency is reintegrated with ecological processes, especially through careful understanding of carrying capacity, preservation and restoration of native diversity and ecosystem health.
- The goal of economic activity is to achieve the highest possible level of cooperative self-reliance.
- Reliance on locally manufactured and maintained appropriate technology, devised through an on-going program of ecological design research, is favored.

A future of promise

Bioregionalism continues to evolve, both as an intellectual and political endeavor. Bioregionalism has taken root in Australia, the United Kingdom, Spain, Italy and Japan, and many other nations. Bioregional periodicals with titles such as *Inhabit*, *ArcoRedes*, and *Lato Selvatico* are successfully extending the cultural range and overall vitality of the bioregional movement.

Bioregionalism is a story best learned from listening to many voices. It is a tale with plot-lines and characters that weave and quickly extend in often

Byzantine interconnection. In attempting to introduce the barest outline of this story I have attempted to be fair to historical fact, and have also tried to introduce sources that allow further exploration of facets of bioregionalism that are all worthy of deeper study. My hope is that key aspects of the breadth and depth of bioregionalism and the bioregional movement, as I see them, have been introduced in a clear, accessible and even inspiring manner. This survey, and others that will follow, will ensure that bioregionalism will no longer be so obscure a notion, and that its concepts can no longer be so easily misappropriated by mainstream government institutions intent on only partial measures of reform.

Bioregionalism is a synthesis of thought, applied technique and persistent practice that is spreading like the patterns of a growing fractal. As people reinhabit their home place, a remarkable integration of philosophy and political activity evolves. Place is perceived as irrevocably connected to culture. Culture is seen as connected to past histories of human and ecosystem exploitation. Constraints to achieving the alternative of a socially-just and ecologically sustainable future are identified, analyzed and confronted. Processes of resistance and renewal are animated within, and parallel to, existing power structures.

To those who hear only a part of the bioregional story, or who attempt to analyze bioregionalism only through the filters of academic or institutional specialties, it may seem to suffer a host of apparent weaknesses, contradictions, or unresolved conflicts. For those who take the time to listen to more of the voices that are speaking about bioregionalism, or better yet participate in the bioregional movement itself, chaos transforms itself into something that is properly perceived as an elegant, persistent and organic growth of purpose. As the human race collectively stumbles into a new millennium, bioregionalism offers the best hope we have for creating an interdependent web of self-reliant, sustainable cultures.

Bibliography

Aberley, D. (1985) "Bioregionalism: A Territorial Approach to Governance and Development of Northwest British Columbia," unpublished MA thesis, University of British Columbia.

—— (ed.) (1993) *Boundaries of Home: Mapping for Local Empowerment*, Gabriola Island: New Society Publishers.

—— (ed.) (1994) *Futures by Design: The Practice of Ecological Planning*, Gabriola Island: New Society Publishers.

Alexander, D. (1990) "Bioregionalism: Science or Sensibility?" *Environmental Ethics* 12 (2): 161–73.

—— (1993) "Bioregionalism: The Need for a Firmer Theoretical Foundation," in J. Vorst *et al.* (eds.) *Green on Red: Evolving Ecological Socialism*, Winnipeg: Society for Socialist Studies.

Andruss, V., Plant, C., Plant, J. and Wright, E. (eds.) (1990) *Home! A Bioregional Reader*, Philadelphia: New Society Publishers.

Atkinson, A. (1992) "The Urban Bioregion as 'Sustainable Development' Paradigm," *Third World Planning Review* 14 (4): 327–54.

Berg, P. (1971) *Homeskin*, San Francisco: Peter Berg.

—— (1976) "Amble Toward Continent Congress," *Continent Congress, Bundle Number 4*, San Francisco: Planet Drum Foundation.

—— (1977) "Strategies for Reinhabiting the Northern California Bioregion," *Seriatim: Journal of Ecotopia*, 1 (3): 2–8.

—— (ed.) (1978) *Reinhabiting a Separate Country: A Bioregional Anthology of Northern California*, San Francisco: Planet Drum Books.

—— (1981) "Devolving Beyond Global Monoculture," *CoEvolution Quarterly* 32: 24–30.

—— (1982) *Figures of Regulation: Guides for Re-Balancing Society with the Biosphere*, San Francisco: Planet Drum Foundation.

—— (1983) "More Than Saving What's Left," *Raise The Stakes* 8: 1–2.

—— (1986) "Growing a Life-Place Politics," *Raise The Stakes* 11: 9–12.

—— (1995) *Discovering Your Life-Place: A First Bioregional Workbook*, San Francisco: Planet Drum Foundation.

Berg, P. and Dasmann, R. (1977) "Reinhabiting California," *The Ecologist* 7 (10): 399–401.

Berg, P., Magilavy, B. and Zuckerman, S. (1990) *A Green City Program for the San Francisco Bay Area and Beyond*, San Francisco: Planet Drum Foundation.

Berg, P. and Tukel, G. (1980) *Renewable Energy and Bioregions: A New Context for Public Policy*, San Francisco: Planet Drum Foundation.

Berman, M. (1981) *The Reenchantment of the World*, Ithaca: Cornell University Press.

Berry, T. (1988) *The Dream of the Earth*, San Francisco: Sierra Club Books.

Berry, W. (1977) *The Unsettling of America: Culture and Agriculture*, San Francisco: Sierra Club Books.

Bookchin, M. (1982) *The Ecology of Freedom: The Emergence and Dissolution of Hierarchy*, Palo Alto: Cheshire Books.

California State Resources Agency (1991) *Memorandum of Understanding: California's Coordinated Regional Strategy to Conserve Biological Diversity*, Sacramento: The Resources Agency of California.

Callenbach, E. (1975) *Ecotopia: The Notebooks and Reports of William Weston*, Berkeley: Banyan Tree Books.

—— (1981) *Ecotopia Emerging*, Berkeley: Banyan Tree Books.

Carr, M. (1990) "Place, Pattern and Politics: The Bioregional Movement of Turtle Island," unpublished MA thesis, York University.

Commission on Navajo Government Development (1991) Status Report June 30, 1991: Submitted to the Navajo Nation Council, Window Rock: Commission on Navajo Government Development.

Crofoot, M. (1987) "A Proposal of Marriage," in A. Hart *et al.* (eds.) *North American Bioregional Congress II Proceedings*, Forestville: Hart Publishing.

Dasmann, R.F. (1965) *The Destruction of California*, New York: Macmillan.

—— (1975) *The Conservation Alternative*, New York: Wiley.

—— (1984) *Environmental Conservation*, Somerset: Wiley.

De Bell, G. (ed.) (1970) *The Environmental Handbook*, New York: Intext.

Devine, M. (1992) "Building Nunavut," *Up Here* July: 18–21.

Dodge, J. (1981) "Living By Life: Some Bioregional Theory and Practice," *CoEvolution Quarterly* 32: 6–12.

Dolcini, M., Fahl-King, C., Fahl-King, D., King, B., Mills, S., Montgomery, T. and Traina, F. (eds.) (1991) *Fourth North American Bioregional Congress*, Alpha Farm: Turtle Island Office.

European Communities: Committee of the Regions (1994) Minutes of the Inaugural Session of the Committee of the Regions for the First Four-Year Term of Office (1994–1998), Meeting of 10 March 1994, Brussels: European Communities.

Fike, M.S. and Kerr, S. (1995) "Making the Links: Why Bioregionalism Needs Ecofeminism," *Alternatives* 21 (2): 22–7.

Forsey, H. (ed.) (1993) *Circles of Strength: Community Alternatives to Alienation*, Gabriola Island: New Society Publishers.

Frenkel, S. (1994) "Old Theories in New Places? Environmental Determinism and Bioregionalism," *Professional Geographer* 46 (3): 289–95.

Friedmann, J. and Weaver, C. (1979) *Territory and Function: The Evolution of Regional Planning*, Berkeley, CA: University of California Press.

Frisco Bay Mussel Group (1977) *Living Here*, San Francisco: Frisco Bay Mussel Group.

Furuseth, O. and Cocklin, C. (1995) "Regional Perspectives on Resource Policy: Implementing Sustainable Management in New Zealand," *Journal of Environmental Planning and Management* 38 (2): 181–200.

Grogen, E. (1990) *Ringolevio: A Life Played for Keeps*, New York: Citadel.

Haenke, D. (1984) *Ecological Politics and Bioregionalism*, Drury: New Life Farm.

Halper, J. (ed.) (1991) *Gary Snyder: Dimensions of a Life*, San Francisco: Sierra Club Books.

Harrington, S. (ed.) (1995) *Giving the Land a Voice: Mapping Our Home Places*, Salt Spring Island: Salt Spring Island Community Services Society.

Hart, A., Rehbock, J.-T., Froelich, J., Zuckerman, S. and Montgomery, M. (eds.) (1987) *North American Bioregional Congress II Proceedings*, Forestville: Hart Publishing.

Helm, M. (1986) "Dwellers in the Land: The Bioregional Vision," *Raise The Stakes* 11: 12.

Henderson, D., Steinwachs, M., Haenke, D. and Wittenberg, V. (eds.) (1984) *North American Bioregional Congress Proceedings*, Drury: New Life Farm.

House, F. (1974) "Totem Salmon," *North Pacific Rim Alive, Bundle Number 3*, San Francisco: Planet Drum Foundation; repr. V. Andruss *et al.* (eds.) *Home! A Bioregional Reader*, Philadelphia: New Society Publishers, 1990.

—— (1990) "To Learn the Things We Need to Know: Engaging the Particulars of the Planet's Recovery," *Whole Earth Review* 66: 36–47.

House, F. and Gorsline, J. (1974) "Future Primitive," *North Pacific Rim Alive, Bundle Number 3*, San Francisco: Planet Drum Foundation.

Kansas Area Watershed Council (1982) *KAW Council Resolutions*, Lawrence: Kansas Area Watershed Council.

LaChapelle, D. (1988) *Sacred Land, Sacred Sex: Rapture of the Deep*, 1st edn., Silverton: Finn Hill Arts.

Lawless, G. (1977) *The Gulf of Maine: Blackberry Reader One*, Brunswick: Blackberry.

—— (1994) *Poems for the Wild Earth*, Nobleboro: Blackberry.

Marshall, G. (ed.) (1989) *Proceedings: First Bioregional Congress of the Upper Blackland Prairie*, Dallas: Realistic Living.

Martien, J. (1982) *Groundhog Manifesto*, Trinidad: Jerry Martien.

—— (1984) *The Rocks Along the Coast*, Westhaven: Jerry Martien.

McTaggart, W.D. (1993) "Bioregionalism and Regional Geography: Place, People and Networks," *The Canadian Geographer* 37 (4): 307–19.

Merchant, C. (1992) *Radical Ecology: The Search for a Livable World*, New York: Routledge.

Meyer, C. and Moosang, F. (eds.) (1992) *Living with the Land: Communities Restoring
the Earth*, Gabriola Island: New Society Publishers.

Milbrath, L.W. (1989) *Envisioning a Sustainable Society: Learning Our Way Out*, Albany:
State University of New York Press.

Mills, S. (1989) *Whatever Happened to Ecology?*, San Francisco: Sierra Club Books.

Mollison, B. (1990) *Permaculture: A Practical Guide to a Sustainable Future*, Washington, D.C.: Island Press.

Muller, M. (1984) "Bioregionalism/Western Culture/Women," *Raise The Stakes* 10: 2.

Noble, E. (1997) *The Digger Archives: San Francisco Diggers (1966–1968) . . . and
Beyond* <http://www.webcom.com/~enoble/welcome.html>.

Parsons, J.J. (1985) "On 'Bioregionalism' and 'Watershed Consciousness'," *The Professional Geographer* 37 (1): 1–6.

Payne, L. (ed.) (1992) *Turtle Island Bioregional Congress V: Proceedings*, Texas: Turtle
Island Office.

Pepper, D. (1993) *Eco-Socialism: From Deep Ecology to Social Justice*, London: Routledge.

Plant, C. and Plant, J. (eds.) (1990) *Turtle Talk: Voices for a Sustainable Future*, Philadelphia: New Society Publishers.

—— (1991) *Green Business: Hope or Hoax. Toward an Authentic Strategy for Restoring
the Earth*, Gabriola Island: New Society Publishers.

Plant, J. (1986) "The Power of an Image: Bioregionalism and Eco-Feminism," type-
script, Lillooet, BC: Judith Plant.

Plant, J. and Plant, C. (eds.) (1992) *Putting Power in its Place: Create Community
Control!* Gabriola Island: New Society Publishers.

Rifkin, J. (1991) *Biosphere Politics: A New Consciousness for a New Century*, New York:
Crown.

Robertson, W.A. (1993) "New Zealand's new legislation for sustainable resource
management: The Resource Management Act 1991," *Land Use Policy*: 303–11.

Roszak, T. (1969) *The Making of a Counterculture: Reflections on the Technocratic Society
and Its Youthful Opposition*, New York: Anchor Doubleday.

—— (1975) *Unfinished Animal: The Aquarian Frontier and the Evolution of Consciousness*, New York: Harper & Row.

Royal Commission on the Future of the Toronto Waterfront (1992) Regeneration:
Toronto's Waterfront and the Sustainable City. Final Report, Toronto: Queen's
Printer of Ontario.

Sale, K. (1985) *Dwellers in the Land: The Bioregional Vision*, San Francisco: Sierra Club
Books.

Scott, L. and Carpenter, J. (eds.) (1988) *The Second Bioregional Congress of Pacific
Cascadia: Proceedings, Resources and Directory*, Portland: Bioregional Congress of
Pacific Cascadia.

Seed, J., Macy, J., Fleming, P. and Naess, A. (1988) *Thinking Like a Mountain: Towards
a Council of All Beings*, Philadelphia: New Society Publishers.

Snyder, G. (1969) *Earth House Hold*, New York: New Directions Books.

—— (1974) *Turtle Island*, New York: New Directions Books.

—— (1977) *The Old Ways: Six Essays*, San Francisco: City Lights Books.

—— (1980) *The Real Work: Interviews and Talks 1964–1979*, New York: New Directions
Books.

—— (1983) *Axe Handles*, San Francisco: North Point Press.

—— (1990) *The Practice of the Wild*, San Francisco: North Point Press.

—— (1995) *A Place in Space: Ethics, Aesthetics and Watersheds*, Washington, D.C.: Counterpoint.

Starhawk (1987) *Truth or Dare: Encounters with Power, Authority and Mystery*, San Francisco: Harper & Row.

Sussman, C. (ed.) (1976) *Planning the Fourth Migration: The Neglected Vision of the Regional Planning Association of America*, Cambridge: MIT Press.

Todd, J. and Tukel, G. (1981) *Reinhabiting Cities and Towns: Designing for Sustainability*, San Francisco: Planet Drum Foundation.

Tukel, G. (1982) *Toward a Bioregional Model: Clearing Ground for Watershed Planning*, San Francisco: Planet Drum Foundation.

Van Newkirk, A. (1975) *Institute for Bioregional Research*, Hetherton: Institute for Bioregional Research.

Wackernagel, M. and Rees, W. (1996) *Our Ecological Footprint: Reducing Human Impact on the Earth*, Gabriola Island: New Society Publishers.

Wright, J.C. (1990) *Natural Resource Accounting*, Canterbury: Centre for Resource Management.

Zuckerman, S. (ed.) (1989) *Third North American Bioregional Congress*, Wheeling: North American Bioregional Congress.

3 Place

Thinking about bioregional history

Dan Flores

As he told the story, Walter Prescott Webb – widely accepted among American environmental historians as one of the founding fathers of this exciting new discipline – began to conceptualize his most famous work, *The Great Plains: A Study in Institutions and Environment*, at the age of five. Raised as a young child in Panola County, Texas, deep in the heart of Southern culture, where thickly-timbered rolling hills screened the horizon and even the overhead sky was only partially visible through the soaring loblolly pines, Webb had not yet started school when his family moved to Central Texas. In the Western Cross Timbers province at the edge of the Great Plains, the future thinker of big ideas found himself stimulated by another world. Here no loblolly pines blocked the skies, and across the grasslands the horizon was miles distant, visible like the encircling rim of a bowl in every direction. Here King Cotton and backwoods truck-garden farms gave way to fenced spreads enclosing the Sacred Cow. Young Walter Webb was fascinated at the differences in those two worlds, remained fascinated as an adult, and with his books and articles went on to stimulate reading Americans into pondering the peculiar dialogue between the Western environmental setting and human technological adaptations to it (Tobin 1988; Webb 1931).

Webb's approach to history – like Frederick Jackson Turner's frontier thesis (which Webb claimed not to have read prior to writing *The Great Plains*) – has been attacked over the years. Fred Shannon's 200-page savaging of Webb's book in 1940 was the first of those critiques, made all the more famous by Webb's laconic refusal to acknowledge Shannon's points: his response to Shannon's critique was that he had conceived and written *The Great Plains* not as history, but as art! In our own time, even his sympathizers acknowledge that Webb's basic ideas were an exercise in environmental determinism, an approach that geographers and anthropologists had long since abandoned. Somewhat in the manner of an intellectual Ulysses S. Grant, it has been asserted, Webb was moved to write big idea books like *The Great Plains* and later *The Great Frontier* (1964) essentially because he lacked the education to know better.

I disagree: I think Webb was so moved because as a child he paid attention to what his senses told him about the difference between the Piney Woods and the Western Cross Timbers in Texas. Just as Frederick Jackson Turner's

essentially Darwinian idea that new environmental settings ("frontiers") transformed those peoples who experienced them into new peoples (Americans in
Turner's thinking), Webb's intellectual vision also continues to influence environmental historians, especially environmental historians of the American West.[1]
Donald Worster (1992), for example, has pointed out that Webb's insistence
that aridity is a defining characteristic of the American West has taught us
correctly that "the West" was not a process, but was and is a *region* whose
perimeter can be sensed on the ground, and marked out on a map. Beyond
that, Webb's approach remains valuable in environmental history because of the
attention he forces us to pay to the confluence between specific ecological realities and specific human adaptations (in the relatively simple terms of *The Great
Plains*, these include the use of windmills and barbed wire in semi-arid, open
grasslands) that are a part of the evolution of cultures in place.

Having struggled with terms like *region* and *place* in writing a book about
the Llano Estacado country of the American Southwest, I think I would argue
that the intuitive foundation that Webb built – albeit in a refined form – ought
to remain central to the way we think about ourselves in the context of history
(Flores 1990). Indeed, it seems to me that the particularism of distinctive *places*
fashioned by human culture's peculiar and fascinating interpenetration with all
the vagaries of topography, climate, and evolving ecology that define landscapes
– and the continuing existence of such *places* despite the homogenizing forces
of the modern world ought to cause us to realize that one of the most insightful
ways for us to think about the human past is in the form of what might be
called *bioregional* histories. Although at first glance it might seem counter to
modern appearances, even urban areas sprang into existence, and most often
continue to depend, on environmental circumstances that lie just below the
level of our awareness. From the time humans located regularly-visited hunting
camps and early river valley farming settlements, human places have been super-
imposed on environmental settings. They still are.

The American West is by no means the only region where the insights of
bioregional history seem to be especially valuable, but because it is my field of
study most of the examples here tend to be drawn from there. For instance,
towards the end of one of Patricia Limerick's essays in *Trails: Toward a New
Western History*, she recounts a conversation after one of her public lectures
with a man who said something like "I enjoyed your speech, but since I'm not a
Western historian, everything you said was obvious to me" (1991: 72). At the
University of Montana, the graduate students in history and environmental
studies tend to be young, bright nonresidents seeking exposure to the
Mountain West. Many of them take advantage of their location to travel widely,
from Montana south to Texas, and sometimes northward through Canada to
Alaska. I suspect this is why they are as puzzled as Patricia Limerick's acquain-
tance when they read historical essays positing various reasons why the
American West as a whole comprises a distinctive, singular region much as the
American South appears to do. Their questions often follow these lines:

"If aridity is the defining characteristic of the West, does it mean that neither Alaska nor the Pacific Northwest belong? and that those high, wet lifezones disqualify the Rocky Mountains?"

"Isn't it ironic that Texas has produced some of the most potent Western symbols, but isn't really a part of the West because it lacks the defining system of federal land ownership?"

"To me, Colorado seems so different from Utah. Are you telling us that they were shaped into a consistent form by the same forces of global economic integration that forged the rest of the West?"

Uncorrupted by an impulse towards the broadly inclusive and generalized definitions of regionalism that professional historians have been trained to apply, these students see the obvious. And of course they are not alone. I have noticed with interest the testimonies of Westerners recorded in the anthology, *A Society to Match the Scenery*, assembled at the Center of the American West at the University of Colorado in 1991 as an exercise in envisioning the future of the West. Among the voices appearing in that anthology were Dan Kemmis (1991) and Camille Guerin-Gonzalez (1991), both residents of the Rocky Mountains (Western Montana and Northern New Mexico), both inhabitants of similar topographies where federal land ownership and management are everyday facts of life, where resource extraction and tourism prevail economically, where water problems and an influx of wealthy newcomers dominate local discussion. Yet after listening to Kemmis' remarks about life in the Northern Rockies, Guerin-Gonzalez claimed that the Northern New Mexico she knows bears no relationship whatsoever to what Kemmis described.

The answer to the puzzlement and to the denials of uniformity expressed above is clearly that in the case of the American West, no set of generalized definitions, regardless of how inclusive, accurately explains the loose cluster of subregions comprising the huge swath of continental topography and ecology that is the Western United States. Neither aridity and its effects, nor federal land ownership, nor economic integration into the global market at a time of mature industrial development, nor the presence of Indian reservations, nor proximity to Mexico or to the Pacific Ocean, nor a legacy of conquest, captures the particularism that is the historical reality of *place* in the Western US.[2] I doubt that broadly-generalized interpretations work very well at capturing historical sense of place elsewhere. But as it has done for a century now, the country west of the Mississippi River continues to work extraordinarily well as a laboratory in humanities and ecology.

Let me here define my own terms and explain my argument by following three lines of investigation. First, if we grant that specific human cultures and specific landscapes can and do, in fact, intertwine to create distinctive places, then why not just turn to careful readings of the local county histories that amateur historians have generated in industrial quantities and that presently fill so much space in local libraries? Given that thousands of such local histories already exist – admittedly, often done uncritically as a kind of pioneer family ancestor worship – what constitutes the rationale and the basis for a distinct human "bioregional history"? Second, if the theories promulgated half a century ago by Walter Prescott Webb and others of his generation (like James Malin) are passé and too obscure and eccentric to serve as ways to think about history and place, what kinds of approaches and ideas ought we to look for in environmental histories of human places, in the American West or anywhere else? Finally, are there works that already exist demonstrating the value of bioregional histories of place in the United States?

On a point of vocabulary, a defense of sorts may be necessary. To some traditionalist academics, my use of the terms "bioregion," "bioregional" and "bioregionalism" may appear to be either an unnecessary resort to jargon, or a surrender to faddism. On the contrary, I would assert that for all its association in the US with countercultural environmentalism, "bioregion" should in fact be recognized as *a precise and highly useful term.* As Doug Aberley showed in Chapter 2, bioregionalism is a contemporary social movement, and should be interesting to environmental historians in its own right. But, it is not merely bioregionalism's focus on ecology and geography, but its emphasis on the close linkage between ecological locale and human culture, its implication that in a variety of ways humans not only alter environments but also adapt to them, that ties it to some central questions of environmental history. While the history of politics and diplomacy and (sometimes) ideas may be extracted from the environmental setting and studied profitably, the kinds of subjects that attract contemporary historical study – legal, social, gender, ethnic, science, technology and environmental issues – require sophisticated reference to places, and to what people think about places. But there is an irony here. Professional history, especially in the US, has long regarded an interest in place as limiting, provincial and probably antiquarian. It is far too common to hear from American historians that those who have spent their careers writing about locales, states, or regions are routinely dismissed as provincial, unless their interest has been New England or California; distressingly often they are dismissed by those who see themselves as writing about more universally crucial (but no less geographically specific) locales like Concord or Languedoc or Washington, D.C..

Environmental histories of place have already made progress against this sniffing condescension, and if as historians who are writing such works we are good at what we do, we are likely to wipe out such attitudes altogether. In truth, to an extent all history is the history of place. But environmental history has gone beyond traditional history and justified its reputation for new insight by following the lead of ecologists, geographers, ecological anthropologists – and bioregionalists – in drawing the boundaries of some of the places we have studied in ways that make real sense ecologically and topographically.

It ought to be agreed that, with rare exceptions, the politically-derived boundaries of county, state and nation are mostly useless in understanding nature. History continues to rely heavily on the trail of documents generated by political life, but much environmental history has recognized that there are significant limitations inherent in that dependency. The founders of American environmental history like Webb and James Malin realized this, and pointed the way towards a more ecologically-oriented kind of study more than half a century ago.[3] Clearly, the first step in recognizing the value of environmental histories of place is a recognition that natural geographic systems (ecoregions, biotic provinces, physiographic provinces, biomes, ecosystems – in short, larger and smaller representations of what we probably ought to call *bioregions*) are the appropriate settings for insightful local history.

Without county/province/state/national borders to provide clues for delimiting place history, to what sources should historians and interested residents turn for ideas about drawing natural boundaries? For the American West, one of the best is the earliest. In 1890, John Wesley Powell laid before Congressional committee a remarkable and beautiful colored map of the arid West, mapping out twenty-four major natural provinces, and further subdividing the region into some 140 candidates for "commonwealth" status based on drainage and topographical cohesion (US Geological Survey 1891). Powell's ideas for the West conform to most twentieth-century delineations, such as those in Wallace Atwood's *The Physiographic Provinces of North America* (1940), and indicate in startling fashion the century-old genesis of the current vogue among Western resource bureaus for ecosystems management.

More ecologically precise and recent than Powell, however, is Robert Bailey's *Ecoregions of the United States* (first published in 1976 and newly revised in 1996) and its accompanying map. Assembled by the US Department of Agriculture from a broad range of soil, climate, floral, faunal and topographic sources, but based primarily on plant study, this work creates a taxonomy of North American nature ranging from the macro (five "Domains" further differentiated into twelve "Divisions") to the micro (the divisions further refined into thirty-one "Provinces," which are themselves subdivided into forty-five "Sections"). The most refined, the sections, range considerably in size. Bailey's "Southeastern Mixed Forest Section," for example, extends across parts of nine states; his "Grama–Buffalo Grass Section" covers all of the southern and central Great Plains. On the other hand, the superficially similar Rocky Mountain region is conceptualized in Bailey as being comprised of eight ecologically unique sections. His system does not specifically locate and provide boundaries for ecosystem corridors such as major rivers and their drainages, which have demonstrably played key roles in human history. But in his taxonomy, somewhere historians might be tempted to view as a single area (like the Intermountain Great Basin), is subdivided into no less than five distinct sections (Bailey 1976).

Typically, scholars closer to the local ground tend to subdivide to the even more specific. Today, in my home states of Montana and Texas, for example, bioregional particularism is rife. In Rocky Mountain Montana, Bailey's two sections are carved by state ecologists into twice that number of categories: the Columbian Rockies, the Broad Valley Rockies, the Yellowstone Rockies and the Rocky Mountain Foreland. More recently, ecosystems ecologists in

Montana/Wyoming have recognized the dynamism of bioregions, and the role played by cultural developments, with their designation of two modern natural systems they call the Greater Yellowstone Ecosystem and the Northern Continental Divide Ecosystem. As for Texas, the majority of ecologists see the state as an extraordinarily artificial creation that cobbles together no fewer than *ten* individual bioregions (one more than Bailey recognizes).[4] Beyond providing scholars with a new axle upon which to spin environmental history, bioregional study at least partially explains why a "place" like Texas cannot decide whether it is properly Southern, Western, Southwestern, or just Texan.

The particularism of bioregional places that is often so observable to travelers is explained by the geographer Yi-Fu Tuan's equation: space plus culture equals place (1977: 4–6). Beyond ecological parameters, the second basis for thinking in terms of bioregional history, of course, is the existence of a diversity of human cultures across both time and space. Since I began this essay by invoking Webb, it is worth mentioning as a basic framework for understanding the dialogue between nature and culture that Webb was properly criticized in his day for his resort to environmental determinism in writing *The Great Plains*. As he saw it and wrote it, certain characteristics of nature on the Great Plains presented so many challenges to American settlement culture as it had evolved in the eastern woodlands that the Great Plains represented an "institutional fault line" that significantly modified the settlement strategies (by which Webb meant primarily materialist economic culture) of the peoples who moved there. In something of a major misreading of history, Webb believed that Hispanic culture had failed to adapt to the Plains to the same extent that horse Indians and Texans did, and that this adaptive failure explained why more technologically innovative Anglo-Americans seized the region (Webb 1931: 85–139).

James Malin, among other historians, corrected Webb's naive assumptions about environmental determinism with his application of the interpretive framework of "possibilism" (Malin 1984). Although Malin still accepted a dichotomy between nature and culture and had an inordinate faith in technological solutions, in possibilism human cultures bear a sturdier freight and responsibility in creating places. The possibilist idea is well-understood now to imply that a given bioregion and its resources offer a range of possibilities, from which a given human culture makes economic and lifeway choices based upon the culture's technological ability plus its ideological vision of how the landscape is seen and ought to be shaped and used to meet a given culture's definition of a good life. While the possibilist idea is scarcely new, and in the social sciences has long been retired from the cutting edge of interpretation, I happen to believe that, carefully used, the idea has a continuing relevance for thinking about environmental history, where interest in the peculiar wake of events that follows ideologies and choices through time is particularly keen.

Resting culture's role in bioregional history on a possibilist model was a reaction to environmental determinism. It continued the nature/culture dichotomy and carried with it the danger of playing to modernism's conceit that ever since the scientific and industrial revolutions, human culture had triumphed and nature hardly mattered anymore except as a potential commodity. Possibilism alone can give the impression – and it is a frequent failing of the bioregional movement's own philosophy – that historical decisions about place are formed exclusively by local populations. Springing particularly from ecologist Eugene Odum's influential studies of ecosystems, since the early 1970s scholars of culture have had a set of mechanisms known collectively as systems theory to explain the diverse web of connections that tie local places to diverse economic and ideological systems.[5] Since Odum, one sort of footbridge has been constructed to span the intellectual river between nature and culture: Not only biological processes but cultural ones appeared to operate as functional, and evolving, systems. Some of the best recent environmental histories – Donald Worster's *Dust Bowl* (1982), Richard White's *Roots of Dependency* (1983), William Cronon's *Nature's Metropolis* (1992) – have made the concept of systems central to their analysis, primarily (following the lead of anthropologists like Marvin Harris) in a materialist and market-integrative form. Hence, Worster blames the Dust Bowl not on farmers but on the "system" of capitalism in a fragile place; White shows how the global market system ensnared Indians so that eventually their ecologies broke down; and Cronon shows how an urban area like Chicago created a system that exploited everything from soil to buffalo hundreds of miles distant.

One remaining element of cultural study that has relevance to bioregional history is the recent refinement of cultural adaptation theory. Inspired by the Odum-derived thought that human systems might, after all, be living organisms, in the late 1970s social scientists like Karl Butzer and Roy Rappaport applied the precepts of organic evolution to cultural adaptation, and tried to sort out its mechanisms in various simulations of real life, as in the feedback loops of a land-management bureaucracy dealing with an environmental crisis, for example. While Butzer (1980) wonders whether entire human civilizations might not *be* organisms, Rappaport (1977) asserts that ultimately adaptation's function is the same whether it occurs in organisms or societies. That function is to aid survival. "Since survival is nothing if not biological," Rappaport wrote, "evolutionary changes perpetuating economic or political institutions at the expense of the biological well-being of man, societies and ecosystems may be considered maladaptive" (1977: 69–71). In Rappaport's view, then, adaptation is critical to understanding the long-term successes (or the short-term failures) of human cultures in specific places, and positive adaptations are those that intertwine cultural choices with the dynamism of particular bioregions into a mix that "survives."

For the resident of a particular place, thinking about local bioregional history has to commence with the recognition that one's place exists at least in part on the taxonomy I've outlined above, although given the natural human preference for ecotone edges, interesting settings for human history won't necessarily be bounded the way Bailey maps out his ecoregions. Pre-Columbian Indian cultures in North America hewed fairly closely to the larger bioregional divisions we now recognize, and individual bands often conformed in their home ranges roughly to slices of topography that Bailey has identified in his sections. On the other hand, Western cultural groups like the New Mexican Hispanics and the Mormons occupied and adapted to several kinds of bioregions in the American West. All manner of intriguing possibilities exist when culturally distinctive groups like Indian tribes, Hutterites, or Mennonites emerge, island-like, in seas of cultural homogeneity like the Great Plains, or when the political boundaries of different traditions cut across a bioregion with its own historical arc.[6]

In an article in *History News*, Hal Rothman (1993: 8–9) has expressed concern that histories of place might succumb to provincial bias if they are written about places to which the authors have an emotional tie. That *could* be a danger for someone whose connections are to local culture, who might take offense at environmental history's tendency to set biodiversity, say, on an equal basis with human economic success. In the books that do exist in this field, however, I would assert that one would have to look very hard to find much native Kansan sympathy (although his family springs from a Kansan background) in Worster's *Dust Bowl*; the author's sympathy in this work tends to be a broader-based identification with what he presents as a healthier landscape as it existed during Native American tenure. My own experience in writing bioregional history is similar: An emotional tie to a landscape made me considerably more critical of some of the human cultures occupying the Llano Estacado than a detached objectivity might have done. On the other hand, the best writing and most penetrating research consistently spring from passion, and places can summon that.

Anyone looking for a sophisticated modern environmental history of place is probably going to have to make peace with a current view of a natural world that is dynamic. Far from serving as some pristine baseline of climax harmony, our bioregions as presented by the new ecology have to be accepted as endlessly evolving through time. Ecologists now speak of "internal change," "blurred secessional patchworks," "moving mosaics." Disturbance is the natural state, hence adjustment is on-going and fundamental. There has been some resistance to this among historians who have hoped, in the environmentalist tradition, that there *is* a harmonious, stable nature out there against which we might view human activity as arrayed in a destructive assault.[7] But while Daniel Botkin's view, that "Nature undisturbed [by human activity] is not constant in form, structure, or proportion, but changes at every scale of time and space" (1990: 62) might be problematic for environmental romanticism, I don't see it so for history. In fact, recognizing that the ground of the natural world is shifting – and always has – can be another bridge tying the activity of human culture back into nature.

If they make any claim to apprehending reality, bioregional histories should capture this changing character of place. Indeed, in modern techniques like repeat photography, fine-resolution remotely-sensed data and the manipulation of spatial information with computers (historic maps I have seen recently showing vegetation and fire patterns in the Northern Rockies come to mind), we have the ability to track those changes at a denser grain than ever before. Personally, I do not in any case see how – as either writers or readers of bioregional histories – acknowledging the fact of ongoing, natural disturbance in nature prevents us from critiquing human disturbances that were foolish in an anthropocentric sense, or reprehensible from the perspective of the diversity of life.

Perhaps an element that distinguishes bioregional history from traditional histories, even of places, is a precise spatial application of Fernand Braudel's *longue durée*. For proper perspective, good bioregional history ought to aim for the "big view" not so much through wide geographic generalizations in shallow time, but through analyzing deep time in a single place. As I attempted to show in a recent piece interpreting nineteenth-century Indian environmental history in the West, an accurate understanding of shallow time often isn't possible without the context of the *longue durée*.[8] Bioregional histories, then, properly commence with geology and landform, then take up climate history (again using an array of modern approaches) from ice cores and pollen analysis to packrat middens and dendrochronology. Climate has always been and remains one of the most visible forces interacting with human history. A climate record of place can then position us to understand the ebb and flow of floral and faunal species across space and time the way our eyes enable us to track cumulus clouds drifting across an open basin by the shadows they cast on the ground.

When we analyze human culture in our stories of place – even if our search is for clues, for instance, as to how earlier cultures in a place coped with a warmer climatic regime – I suspect we would do better to cease our quest for Golden Age utopias. Although there are past cultures, no question, that saw and utilized nature differently than modernism does, the study of bioregions indicates that human stress on nature is very ancient. Superficially, bioregional history might seem to confirm our suspicions that the ancient ways were morally superior; closer examinations, however, usually turn out to be more sobering. Further, at every level of time (including our own) we ought to recognize that the supposed dichotomy between culture and nature is not structurally basic to human consciousness, but is a false dichotomy. The preliminary studies in biophilia and biophobia research, for example, indicate in rather striking ways that genetically it is ludicrous to think that humans ever stepped outside nature. Evolutionary psychology and sociobiology, for instance, remind us how rooted our social behavior is in the primate world. And studies of inherited biophobic responses (to snakes and spiders, for example) as well as genetically-transmitted biophilic preferences (to savannahs, parklands, certain tree shapes and terrain scales) seem to center human fear of the natural world, as well as human settlement strategies and even aesthetics, in adaptations selected by evolution over deep time.[9] The extraordinary range that human cultures have taken around the world and throughout time may make such a statement seem counter-intuitive, but we remain children of nature. Unless we are so strongly socialized to ignore our natural surroundings that we take no cognizance of them at all, our cultures endlessly react to the natural world around us, even if only to fear or dread it.

When human cultures in specific places do address nature directly, those ideas can be seen as adaptive packages of "captured knowledge" about living in particular places.[10] Our goal, then, should be to think about local and regional history in a way that sees human cultural adaptation and knowledge transmission essentially as analogous to the natural selection of characteristics. As an example of how this might work in a historical case in the American West, geographer William Riebsame (1991) has recently melded some of the ideas of systems analysis and adaptation theory to distinguish between positive adaptation (a culture's willingness to change in the face of new circumstances) and cultural resiliency (a system's resistance to change, and if perturbed to return rapidly to its former condition). In Riebsame's view, the Southern Plains' response to the Dust Bowl of the 1930s is a classic instance of resiliency rather than adaptation. Rather than a wholesale rethinking of the premise of plowing up the grasslands to create wheat and cotton farms, the regional and national response to the Dust Bowl represented a tinkering with the existing system. While that tinkering (in the form of new agronomy techniques and a small-scale return of marginal farmland to grass) was seen by residents as a successful adaptation, Riebsame thinks that over the long-term, with continuing droughts and collapses, those actions will be re-interpreted as having been a resilient rebound to the status quo, and, hence, ultimately maladaptive.

The narrative line of bioregional history is essentially imagining the stories of different but sequential cultures occupying the same space, and creating their own succession of "places" on the same piece of ground. Because this idea provides cause for so many observable effects in environmental history, it is important to realize in this kind of history that successive cultures inhabiting a space interact with a "nature" more or less altered by the previous inhabitants.[11] Further, we ought to understand that the structure of the dialogue – and that is the proper way to describe it – between nature and human culture is the same kind of dialogue that exists between habitat and species in natural selection. Human cultures alter their places to shape them in accordance with their ideological visions, and in turn cultures are shaped by the power of their places.[12] Just as we now understand organic evolution, significant changes in bioregional histories can be expected to occur as punctuations in equilibria, rapid ratchetings – or "ecological revolutions" to borrow historian Carolyn Merchant's term (1989: 2–3) – to new conditions that have the spiraling effect of endlessly recreating place. Thinking about the historical causes and results surrounding these ratchets to new conditions forces us to pay close attention to materialist and economic changes, naturally enough. But some of the most penetrating insights spring from studies of ideologies, values, literature, art – the endless and changing ways we humans have *imagined* and portrayed our places.

What separates historical writing from the semantically-challenged language of the social scientist, and makes history rather more useful to the reading public, is history's greater burden to communicate. Unlike geographers, anthropologists and sociologists, historians communicate generalities with stories of individuals, whose experiences carry more of the scent of life for readers.[13] It is exactly the "fuzzy" propensity for anecdote, the discipline's inherent wish to tell stories, that drives history and makes it readable. Like all good writing, quality history "shows": It doesn't tell, and it doesn't seek obfuscation in resort to jargon. As Yi-Fu Tuan wrote two decades ago in *Topophilia*, affection for history, as with place, tends to focus on smaller and more personal scales than the large political boundaries of the modern world, and human sense of place has everything to do with a shared sense of history (1974: 93–112). I take this as confirmation that there is a very eager audience for lovingly-crafted bioregional history.

Bioregional histories of the kind I've described here do not yet exist in quantity, but the list is growing. Despite its county focus and overly-academic title, Richard White's first book, *Land Use, Environment, and Social Change: The Shaping of Island County, Washington* (1980), was a promising start to doing modern bioregional history. It showed well how a small place can encapsulate and exemplify many broader historical themes, and in the Pacific Northwest it is now regarded as basic to bioregional literature.[14] Worster's *Dust Bowl* book *is* bioregionally-centered, and certainly explores adaptation, but since its topic is a specific event, it only superficially examines sequential cultures or deep time. The literature and art it explores are those of event rather than place. To the extent that it examines the Imperial Valley of California especially, the same can be said of Worster's *Rivers of Empire: Water, Aridity, and the Growth of the American West* (1985). These two books are almost indispensable to environmental history, but neither is quite a bioregional history.

William Cronon's *Changes in the Land: Indians, Colonists, and the Ecology of New England* (1983), Carolyn Merchant's *Ecological Revolutions: Nature, Gender, and Science in New England* (1989), Richard Judd's *Common Lands, Common People: The Origins of Conservation in Northern New England* (1997), Albert Cowdrey's *This Land, This South: An Environmental History* (1983), Timothy Silver's *A New Face on the Countryside* (1990) and Mart Stewart's *"What Nature Suffers to Groe": Life, Labor, and Landscape on the Georgia Coast, 1680–1920* (1996) have made the bioregions of the Eastern United States perhaps the best studied on the continent. Cowdrey's book, although it mentions deep time, essentially is a broad geography/shallow time work, far different from Stewart's, which covers a broad stretch of history and is an exceptionally multicultural and sophisticated book. Cronon's, Merchant's and Silver's books are all more temporally focused and do explore processes and changes that create places across cultural lines, although without much reference to adaptation. Of the three New England books, Merchant's is the broadest, but theory and jargon poke through the fabric of the writing. Cronon's and Judd's new book are more readable by wide margins. Finally, Philip Scarpino established a different and useful bioregional category with his *Great River: An Environmental History of the Upper Mississippi, 1890–1950*

(1985), a book that, in fact, is somewhat narrow temporally as well as in its focus on industrial and bureaucratic developments.

Among more recent, tightly-focused books concentrating on bioregions in Western America, geographer Robin Doughty's pair of works, *Wildlife and Man in Texas: Environmental Change and Conservation* (1983) and *At Home in Texas: Early Views of the Land* (1987) can be taken together as a shallow-time place history, primarily of the bioregions of Central Texas, although Doughty addresses only Anglo- and German-American cultures there. Hal Rothman's *On Rims and Ridges: The Los Alamos Area Since 1880* (1992) and Peter Boag's *Environment and Experience: Settlement Culture in Oregon* (1992) are highly place-specific: the Parajito Plateau of New Mexico's Jemez Mountains in Rothman's case and the Calapooian Valley of Oregon in Boag's. Rothman's book, an exploration of growing competition for local resources into modern times, is effectively intercultural; Boag's is less so and is limited temporally to the nineteenth century, but is an interpretively rich and imaginative work. My bioregional book on the Southern High Plains, *Caprock Canyonlands: Journeys into the Heart of the Southern Plains* (1990), attempted an experimental approach to history, but did make an effort to incorporate most of the ideas I've mentioned above. My own choice as the best bioregional history anyone has written to date is William deBuys' *Enchantment and Exploitation: The Life and Hard Times of a New Mexico Mountain Range* (1985). It is place-specific (the Sangre de Cristo Range of New Mexico), temporally deep, examines environmental change across sequential cultures, and deals with values and adaptation with an effortless style.

In its brief three decades, modern environmental history has made a name for itself primarily as a field that has offered stimulating studies of environmentalism as a socio-political movement, of intellectual ideas about nature, and of specific environmental events of historical importance. For its theoretical framework, it has mostly borrowed from elsewhere. Yet in the work of Turner, and particularly Webb and James Malin, there existed from the beginning a focus on places and their history, and at least the rudimentary foundations of how to approach that kind of study. As Malin put it thirty-five years ago, the "proper subjects of study" for a specific bioregion are "its geological history, its ecological history, and the history of human culture since the beginning of occupance by primitive men."[15]

Undoubtedly a serious mistake we historians have made in our work – and naturally we have carried readers along with us, since they've had little choice in the matter – is to ignore and devalue local and regional history in favor of endless histories that present the nation-state, its wars, politics and so forth, as the true measure of what history ought to be. Or, alternatively, histories that attempt to impose upon a diverse world some interpretive framework and then to set about forcing the world around us into some facsimile of that model. A more enlightened and useful kind of history for contemporary readers, it would seem, goes after the reality of the specific, presenting sophisticated, deep time, cross-cultural, environmental histories of places, histories that bring us to think about ourselves as inhabitants of places, of watersheds and topographies, of an

[margin notes, handwritten:]
best bioreg-book

subjects of study

objectives of study

evolving piece of space (with an evolving set of fellow inhabitants) different from every other one.

A new kind of history called bioregional history, in other words.

Acknowledgments

A draft of this essay was first published in *History Environmental Review*. Reprinted by permission.

Editor's note: The author's previously published essay was revised for a general audience.

Notes

1 See Cronon 1987: 157–76. Howard Lamar has observed more recently that within the past decade, a survey of historians of the American West indicated that it was Webb's ideas rather than Turner's that they found most stimulating (see Lamar 1992: 25).

2 To aridity – the causative factor in American Western history for Webb, John Wesley Powell and Wallace Stegner – contemporary historians have added the others I mention in the text. All these are singled out to explain Western homogeneity in the following articles in *Trails*: see, particularly, Limerick's "The Unleashing of the Western Public Intellectual," 70–1; Elliott West, "A Longer, Grimmer, But More Interesting Story," 103–11; Michael Malone, "Beyond the Last Frontier: Toward a New Approach to Western American History," 139–60. Also, Susan Neel, "A Place of Extremes: Nature, History, and the American West," in C.A. Milner II (ed.) *A New Significance: Re-Envisioning the History of the American West*, New York: Oxford University Press, 1996.

3 While Webb argued that his book was about the Great Plains, and offered semi-aridity, treelessness and lack of topographical relief as the defining characteristics of the Plains, many readers have observed that his maps implied that virtually all of North America west of the 98th meridian belonged to the Plains province. However, most of his historical examples came from Texas (see Webb 1931: Map 1).

4 Montana Environmental Quality Council, *Fourth Annual Report*, Helena: State of Montana, 1975. On the more recent ecosystem bioregions in the Northern Rockies, see the chapter "The Yellowstone Ecosystem and an Ethic of Place" in Charles Wilkinson, *The Eagle Bird: Mapping a New West*, New York: Pantheon, 1992: 162–86. On Texas bioregions, see Mike Kingston (ed.) *The Texas Almanac, 1994–95*, Dallas: Dallas Morning News, 1994: 94. In Texas, the tenth bioregion not represented separately in Bailey is the Blackland Prairie.

5 See Eugene Odum, "The Strategy of Ecosystem Development," *Science* 164 (April 1969): 262–70, and *Fundamentals of Ecology*, Philadelphia: W.B. Saunders, 1971. A good overview and introduction to the various systems models devised for human societies – modernization theory, dependency theory, world-systems theory – may be found in Thomas Hall's *Social Change in the Southwest, 1350–1880*, Lawrence: University Press of Kansas, 1985: 11–32.

6 For example, Howard Lamar points out the striking differences in the Canadian and American responses to the Dust Bowl (1992: 25–44). On North American cultural regions, see Alfred Kroeber, *Cultural and Natural Areas of Native North America*, Berkeley: University of California Publications in American Archaeology and Ethnology, 1939; Raymond Gastil, *Culture Regions of the United States*, Seattle:

University of Washington Press, 1975; and Joel Garreau, *The Nine Nations of North America*, New York: Avon, 1992.

7 See especially Donald Worster's three essays, "The Shaky Ground of Sustainable Development," "The Ecology of Order and Chaos" and "Restoring a Natural Order" (1993: 142–83), wherein Worster argues hopefully (p. 181) that ecology "will eventually come back with renewed confidence" to the older models!

8 Dan Flores, "Bison Ecology and Bison Diplomacy: The Southern Plains from 1800 to 1850," *Journal of American History* 78 (September 1991): 465–85. What *longue durée* history implies, of course, is that bioregional historians have a sound grasp of paleontology and archeology, as well as ecology and climate study.

9 On human evolutionary psychology and sociobiology, see Jared Diamond, *The Third Chimpanzee: The Evolution and Future of the Human Animal*, New York: HarperCollins, 1992; Richard Dawkins, *The Selfish Gene*, New York: Oxford University Press, 1976; Diane Ackerman, *A Natural History of the Senses*, New York: Random House, 1990.

   The best recent introduction to biophilia and biophobia is contained in the essays collected in Stephen Kellert and Edward Wilson (eds.) *The Biophilia Hypothesis*, Washington: Island Press, 1993. For my points in the text, see, especially, Kellert's "The Biological Basis for Human Values of Nature," 42–69; Robert Ulrich, "Biophilia, Biophobia and Natural Landscapes," 73–137; and Judith Heerwagen and Gordon Orians, "Humans, Habitats and Aesthetics," 139–172.

   Ulrich (p. 125) concludes that genetic biophilias and biophobias may be 20 to 40 percent determinative, but probably have to be triggered by learning.

10 I derive the term "captured knowledge" from Joel Gunn, "Global Climate and Regional Bio-Cultural Diversity," in Carole L. Crumley (ed.) *Historical Ecology: Cultural Knowledge and Changing Landscapes*, Santa Fe: School of American Research Press, 86–90. For examples of this kind of approach, see the anthology Mark McDonnell and Stewart Pickett (eds.) *Humans as Components of Ecosystems*, New York: Springer, 1993 (with foreword by William Cronon).

11 Bill Cronon has appropriated the term "second nature" to describe these culturally altered settings, but I think I would have to insist that for the last 11,000 years, very few human societies have interacted with anything else (see Cronon 1992: 266–7). On the human shaping of North America before the arrival of Europeans, the best general discussion I have seen is William Denevan, "The Pristine Myth: The Landscapes of the Americas in 1492," *Annals of the Association of American Geographers* 82 (September 1992): 369–85.

12 See Winnifred Gallagher, *The Power of Place: How Our Surroundings Shape Our Thoughts, Emotions, and Actions*, New York: Poseidon, 1993; Dan Flores, "Spirit of Place and the Value of Nature in the American West," *Yellowstone Science* 3 (Spring 1993): 6–10.

13 Bioregional historians who properly seek to bring their work to life by interweaving the stories of individuals should be aware, to quote Amos Hawley, that the "basic assumption of human ecology . . . is that adaptation is a collective rather than an individual process. And that in turn commits the point of view to a macrolevel approach" (1986: 126).

14 Conversation with Richard White, 9 April 1994, Missoula, Montana.

15 The quotation is from Swierenga's introduction (p.129) to Malin's chapter titled "On the Nature of the History of Geographical Area" (Malin 1984: 129–43). Malin goes on in this essay to assert that "The study of the history of the western United States as a geographical area is *not* the study of 17, 20, or 22 separate *states* that lie within that area" (1984: 130; my emphasis).

   I would be remiss if I did not mention that Donald Worster has described something like the history I am calling for in a 1984 essay titled "History as Natural History" (repr. Worster 1993: 30–44).

References

Atwood, W. (1940) *The Physiographic Provinces of North America*, Boston: Ginn & Co.

Bailey, R. (1976) *Ecoregions of the United States*, United States Forest Service, Washington: Department of Agriculture; revised edn., 1996.

Berg, P. (1977) "Strategies for Reinhabiting the Northern California Bioregion," *Seriatim: The Journal of Ecotopia* 3: 2.

Boag, P.G. (1992) *Environment and Experience: Settlement Culture in Nineteenth-Century Oregon*, Berkeley: University of California Press.

Botkin, D. (1990) *Discordant Harmonies: A New Ecology for the Twenty-First Century*, New York: Cambridge University Press.

Butzer, K. (1980) "Civilizations: Organisms or Systems?" *American Scientist* 68 (September–October): 517–24.

Cowdrey, A.E. (1983) *This Land, This South: An Environmental History*, Lexington: University Press of Kentucky.

Cronon, W. (1983) *Changes in the Land: Indians, Colonists, and the Ecology of New England*, New York: Hill & Wang.

—— (1987) "Revisiting the Vanishing Frontier: The Legacy of Frederick Jackson Turner," *Western Historical Quarterly* 18 (2): 157–76.

—— (1992) *Nature's Metropolis: Chicago and the Great West*, New York: W.W. Norton.

DeBuys, W.E. (1985) *Enchantment and Exploitation: The Life and Hard Times of a New Mexico Mountain Range*, Albuquerque: University of New Mexico Press.

Doughty, R.W. (1983) *Wildlife and Man in Texas: Environmental Change and Conservation*, College Station: Texas A&M University Press.

—— (1987) *At Home in Texas: Early Views of the Land*, College Station: Texas A&M University Press.

Flores, D. (1990) *Caprock Canyonlands: Journeys into the Heart of the Southern Plains*, Austin: University of Texas Press.

Guerin-Gonzalez, C. (1991) "Freedom Comes from People, Not Place," in G. Holthaus *et al.* (eds.) *A Society to Match the Scenery: Personal Visions of the Future of the American West*, Niwot: University Press of Colorado.

Hawley, A. (1986) *Human Ecology: A Theoretical Essay*, Chicago: University of Chicago Press.

Jones, S. (1959) "Boundary Concepts in the Setting of Place and Time," *Annals of the Association of American Geographers* 49: 241–55.

Judd, R.W. (1997) *Common Lands, Common People: The Origins of Conservation in Northern New England*, Cambridge: Harvard University Press.

Kemmis, D. (1991) "The Last Best Place: How Hardship and Limits Build Community," in G. Holthaus *et al.* (eds.) *A Society to Match the Scenery: Personal Visions of the Future of the American West*, Niwot: University Press of Colorado.

Lamar, H. (1992) "Regionalism and the Broad Methodological Problem," in G. Lich (ed.) *Regional Studies: The Interplay of Land and People*, College Station: Texas A&M University Press.

Limerick, P.L. (1991) "The Unleashing of the Western Public Intellectual," in P.L. Limerick, C. Milner and C. Rankin (eds.) *Trails: Toward a New Western History*, Lawrence: University Press of Kansas.

Malin, J. (1984) *History and Ecology: Studies of the Grasslands*, Lincoln: University of Nebraska Press.

Merchant, C. (1989) *Ecological Revolutions: Nature, Gender, and Science in New England*, Chapel Hill: University of North Carolina Press.

—— (1989) *Ecological Revolutions: Nature, Gender, and Science in New England*, Chapel Hill: University of North Carolina Press.

Rappaport, R. (1977) "Maladaptation in Social Systems," in J. Friedman and M.J. Rowlands (eds.) *The Evolution of Social Systems*, London: Duckworth.

Riebsame, W. (1991) "Sustainability of the Great Plains in an Uncertain Climate," *Great Plains Research* 1: 133–51.

Rothman, H. (1992) *On Rims and Ridges: The Los Alamos Area Since 1880*, Lincoln: University of Nebraska Press.

—— (1993) "Environmental History and Local History," *History News* 48: 8–9.

Scarpino, P.V. (1985) *Great River: An Environmental History of the Upper Mississippi, 1890–1950*, Columbia: University of Missouri Press.

Shannon, F. (1940) "An Appraisal of Walter P. Webb's *The Great Plains: A Study in Institutions and Environment*," *Critiques of Research in the Social Sciences* III (Bulletin 46).

Silver, T. (1990) *A New Face on the Countryside: Indians, Colonists and Slaves in South Atlantic Forests, 1500–1800*, Cambridge and New York: Cambridge University Press.

Stewart, M. (1996) *"What Nature Suffers to Groe": Life, Labor, and Landscape on the Georgia Coast, 1680–1920*, Athens: University of Georgia Press.

Tobin, G. (1988) "Walter Prescott Webb," in J. Wunder (ed.) *Historians of the American Frontier: A Bio–Bibliographical Sourcebook*, New York: Greenwood Press, 713–29.

Tuan, Y. (1974) *Topophilia: A Study of Environmental Perception, Attitudes, and Values*, Englewood Cliffs: Prentice-Hall.

—— (1977) *Space and Place: The Perspective of Experience*, Minneapolis: University of Minnesota Press.

US Geological Survey (1891) "Arid Region of the United States, Showing Drainage Districts," *Eleventh Annual Report (1889–90): Irrigation Survey, Pt. II*, Washington: Government Printing Office.

Webb, W.P. (1931) *The Great Plains: A Study in Institutions and Environment*, Boston: Ginn & Co.

—— (1964) *The Great Frontier*, intro. by A.J. Toynbee, Austin: University of Texas Press.

White, R.L. (1980) *Land Use, Environment, and Social Change: The Shaping of Island County, Washington*, Seattle: University of Washington Press.

—— (1983) *Roots of Dependency: Subsistence, Environment and Social Change Among the Choctaws, Pawnees and Navajos*, Lincoln: University of Nebraska Press.

Worster, D. (1982) *Dust Bowl: The Southern Plains in the 1930s*, New York: Oxford University Press.

—— (1985) *Rivers of Empire: Water, Aridity, and the Growth of the American West*, New York: Pantheon Books.

—— (1992) "New West, True West," in *Under Western Skies: Nature and History in the American West*, New York: Oxford University Press.

—— (1993) *The Wealth of Nature: Environmental History and the Ecological Imagination*, New York: Oxford University Press, 1993.

—— (1994) *An Unsettled Country: Changing Landscapes of the American West*, Albuquerque: University of New Mexico Press.

Figure 4.1 A sorry state of affairs

Source: Beau Jack McGinnis (1992)

4 Boundary creatures and bounded spaces

Michael Vincent McGinnis

> Life is there, alongside the mind, and the human being is inside the circle this mind turns on, and joined to it by a multitude of fibers.
>
> (Artaud 1993)

Human beings and other animals are boundary creatures. Sky, an ocean's surge, firesides, seeds in waterfalls, and boundaries are part of life and becoming human. In crossing boundaries, I believe a culture can find the seeds of bioregional change and be more complete.

A river's boundaries, for example, extend well beyond its banks. Where I stand, the "circle of life" sends waves of water and Pacific salmon swimming upstream the McNeil (Bear) river. At the river's edge, bears wade through a spring run-off to wait for the salmon's return. Bear cubs imitate their mothers in pursuit of wild salmon. Theirs comes a time when the cub has learned all it can from its mother, and will begin to explore its own fishing techniques and strategies. As native inhabitants of "rivertime," bears and salmon adapt to sensual changes in the river system. The adaptive capacity to respond to this system is based on the animal's ability to mime and learn from others. Adaptation is also based on the animal's ability to sense the subtle changes in the wind, the length of the day, the location of the sun and moon, and the smell in the air. The cubs that survive to adulthood have mimed and learned from their elders, are sensitive to the changes of climate and river flow, and have escaped predators. With each change in the watershed, the animal's ability to mime and adapt is continually put to the test.

As indicators of the reproductive character of a healthy ecosystem, the salmon, bear and people are *boundary creatures* that have coexisted for thousands of years. A boundary creature inhabits more than one world; the salmon, bear and people are linked and are nested "parts" of several distinct but interdependent systems of relationships. Wild salmon can navigate through oceans and fresh water because of their well-developed sensual memory of place. This sense of place drives the salmon deeper into the watershed. From fresh water to the ocean and back to the creek, the sense of place and smell drives the salmon upstream to cross the artificial and natural boundaries that exist along its way. In reproduction, this sense is passed on from one generation to another.

The bear and people re-member and return to the spawning ground of the salmon. For humans, salmon memories are important parts of an oral tradition (oral recollection) and the bioregional history of a place. In northern California, indigenous people refer to the king salmon as "Chief Spring Salmon," "Quartz Nose," "Two Gills on Back," "Lightning Following One Another" and "Three Jumps" (Mills 1995: 152). Each reference to salmon is included in story, dance, ritual and forms of cultural mimesis passed on from one generation to the next. This pre-modern way of knowing, thinking and learning place is bioregional. By other ways and means, the bear also passes on its memories. Human beings can learn from the bear, other animals and places. But the society must be open to the secrets, signs and signals of others and of noble places.

Boundaries of mechanical life

We should recognize that these senses and memories we share with other animals in a community are rapidly fading. Modern institutions make a series of somber choices: to foster the development of formal economies and bureaucracies, to devalue informal economies and diverse communities, to control a "static" nature as a resource, to develop technological and scientific instruments for making exploitation of the environmental machine more effective and efficient (Shiva 1997; Mander and Goldsmith 1996; Brecher and Costello 1994; Sachs 1992). Four values are endemic to mechanical life: bureaucratic organization, economic rationality, modern technology and resource management. These values threaten and endanger place, cultural diversity and the health of the bioregion. The human ecological relationship remains caught in a web of increasing mechanization while the remnants of creative, spontaneous, and social interaction are suppressed for the comfort of modern technology. In a mechanistic society, human organization is essentially problematic, impersonal and disabling

To build a world-machine, modernity continues to transform unique places into mechanized spaces. Identification of one's bioregion is becoming more difficult. The transformation of nature into a machine unfolds with the transformation of humanity into a machine or a cyborg. The definition of cyborg according to *Webster's Ninth New Collegiate Dictionary* is "[*cyb*ernetic + *org*anism] (ca. 1962): a human being who is linked (as for temporary adaptation to a hostile space environment) to one or more mechanical devices upon which some of his vital physiological functions depend." As a hybridized species, the cyborg-self mirrors the mechanization, capitalization and objectification of nature-into-a-machine (Harraway 1985; Bennett 1993; Luke 1996). To build a world-machine, the boundaries between the natural and mechanical are crossed.

At first, this notion of the cyborg sounds strange. Yet, modern humanity's dependence on machines is unequivocal. We enter our machines (car or bus) and quickly become passengers in a world we no longer care to understand. The robot on Mars may be a machine, but its mechanical appendages, its mission and its electronic vision are "remotely" controlled by human beings. Humans *via* the machine are on Mars. As cyborgs, we view the world through the lens of a mechanical and electronic eye (e.g. the computer, television and camera lens). We watch the bombing of Iran on the television. This vision of war is mediated by several machines which separate the deadly impacts of war from the virtual and commercial illusions that are produced by machines. The war "games" that are depicted on the TV screen are virtual illusions which have previously been played out in computer games or on video while our ability to feel the pain of war has been diminished. Umberto Eco writes: "The mass media first convinced us that the imaginary was real, and now they are convincing us that the real is imaginary; and the more reality the TV screen shows us, the more cinematic our everyday world becomes" (1994: 48–9). Real interactive "nature" is becoming imaginary as natural entities depicted on the TV screen go extinct. We view the Yellowstone wilderness through the car window or attempt to capture "Old Faithful" on film. But, the unique smells, dangers and complexity of the Yellowstone ecosystem cannot be captured on film or video. People search for the perfect machine while the loss of place permeates modern society.

As we approach the electronic era, the separation of humanity from place seems inevitable. The conversion of nature-as-machine is the quintessential modern project, and resonates with the historical conversion of "primitive" or "less-developed countries" to civilized, mechanical and industrialized economies. I have witnessed the transformation of my place from a wetland ecosystem to agriculture to *Disneyland*. In each transformation, nature takes on a different meaning. Children are more familiar with a bulldozer or earthmover than the oak and chaparral that were once part of their landscape. The "old" nature/culture relationship becomes part of bioregional history while the new developed "environment" – wildlife theme park, "living museum" and the vacant open space – are geo-graphic images held by a new generation.

One need only look to society's treatment of a free-flowing river. A river's water is perceived as a resource for human use or natural capital. The definition of a resource is "a source of nature redirected for human use" (McGinnis 1994, 1995a). During the last sixty years, this instrumental value of a river has contributed to the development of some 75,000 dams in the US or the literal "rearranging of the waters of the continent" (Palmer 1993: 1). In the politics of hydroelectric power development and irrigation networks, the more-than-economic values of the natural world are silenced. Each dam resonates with the technological treatment of nature as a factory. Drowning Hetch Hetchy to provide power for San Francisco redirects the downhill energy (potential energy being converted in nature into kinetic energy) into paths available for urban use (electric energy). Dams reflect an uncritical social reliance on modern technology, and a form of *spatial apartheid* – each dam separates the unique ecological places (riparian areas and watersheds) to support mechanical yet human developments (irrigation, grain transportation, hydropower and urban development). The dams impound the river and the awe of the spring runoffs is gone. We are pleased with our engineering bureaucrats and visit Bonneville Dam. The river-itself is transformed into "an organic machine, a virtual river"

(White 1995: 108). The virtual river-as-machine can be turned on or off. This is the essential character of a "denatured" river that cyborgic society depends on. As Martin Heidegger writes:

> The original nature that was disclosed and brought to word by the Greeks was later, through two alien powers, *de-natured*. Once through Christianity, whereby nature was, in the first place, depreciated to [the level of] "the create," and at the same time was brought into a relation with super-nature (the realm of grace). Then [it was denatured] through modern natural science, which dissolved nature into the orbit of the mathematical order of world-commerce, industrialization, and in a particular sense, machine technology. . . . We must accordingly leave aside the modern notion of nature, to the extent that we have one in general, where the talk is of streams and waters.
>
> (Heidegger 1980: 195; trans. Foltz 1995: 63)

The denaturing of nature coincides with the dehumanizing effects of large-scale economic development and the construction of the world-machine. Timothy Luke agrees, and argues that "the death of 'the human' unfolds along with the death of 'nature'" (1996: 6). The heart of the artificial world is defended by the denaturing of nature. The cultural ramification of this denaturing is dramatic. A report by the US Department of the Interior (Noss *et al.* 1995) states that every ecosystem in the US is threatened or endangered. Society faces a crisis in education, poverty and homelessness. Our political and economic elite fail to recognize the connection between cultural impoverishment and ecological decay. One out of five children in the US live at the poverty level, and these children live in regions that are in ecological decline. In cities, one out of every four homeless are children. This scenario is being played out worldwide, for material poverty follows technological progress.

The cyborg has more faith in "technique" than connectedness to govern society. This faith in technique encourages the denial of responsibility to place. To deny, according to *Webster's New Collegiate Dictionary*, is "to disclaim connection with or responsibility for: DISAVOW." As individuals and as members of institutions, we deny the importance of place in shaping culture and society. As a machine, the bioregion is viewed as a bundle of natural resources to be managed by the best modern technologies available, and in accordance with the canons of efficiency and effectiveness. Identification with the bioregion is re-placed by possession of a Sierra Club *MasterCard*.

We draw, categorize and order life in accordance to values for bureaucratic organization, economic rationality, modern technology and natural resource management. These values are shown in Figure 4.2.

Bureaucracy is defined by Max Weber as a means of "transforming social action into rationally organized action" (1973: 337). This type of rational control serves the capitalistic state because it represents a powerful means to order society and nature on a massive scale (Hummel 1994). In order to

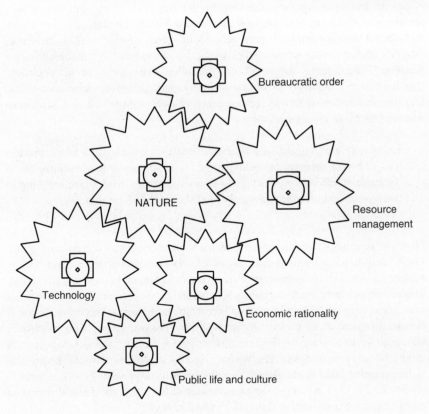

Figure 4.2 The social construction of nature as a machine

control the natural world, bureaucratic organization must construct a view of nature and humanity as a machine so that it can be made more predictable and efficient. The bureaucratic view of nature and humanity as a resource is the essence of economic rationality. Economic rationality opposes an organic life-producing view of natural systems and favors a vision of nature-as-machine that is economy-producing. The machine lacks self-sustaining, sacramental and life-giving qualities.

Economic rationality and the ordering of nature for use is depicted in bureaucratic language. To deal with a fellow predator, technocratic cyborgs refer to the wolf in terms of "predator control," "game unit," "management tools," "control actions," "reduction strategy" and systematically propose to exterminate the wolf. The wolf is viewed as "standing-reserve" [*das Bestand*] to be managed and technologically ordered for use and controlled (see Heidegger 1977: 17–26; Foltz 1995). Viewed as an object, the wolf loses its character as an ecological being; the wolf has no ecological significance in itself. Lacking in the bureaucratic language is an understanding of the predator as an important player in diverse ecosystems. Lacking in the language of the bureaucrat is a recognition of the needs of the wolf in relation to human beings.

Cultural relationships are also defined in terms of the mechanical sense. Records of our economic transactions (e.g. by the use of the "automatic teller machine," Mastercard, or a call on the telephone) are kept in electronic databases, which enhance the ability of the market and government to control, regulate and police society as standing reserve (*das Bestand*). Such a sentiment is shared by Mark Poster who writes:

> Individuals are plugged into circuits of their own panoptic control, making a mockery of theories of social action . . . which privilege consciousness as the basis of self-interpretation, and liberals generally, who locate meaning in the intimate, subjective recesses behind the shield of the skin.
>
> (1995: 87)

Human "resources" are categorized by such technologies as fingerprinting and DNA identification. A human resource is a source of community and society redirected for use by corporate bureaucracies. Human beings are identified, ranked, represented and ultimately controlled as human capital by "human resource managers." The ultimate danger in the ordering of humanity for use is the dehumanization of society. Weber warned: "No one knows who will live in this cage in the future. . . . For the last stage of this cultural development, it might be truly said: Specialists without spirit, sensualists without heart; this nullity imagines that it has attained a level of civilization never before achieved" (1958: 182). We have constructed machines to mirror a mechanical image of nature. And we now inhabit these mechanical cages.

One need only look to the development of the world-city to see how these values are taking hold. In a critical examination of Los Angeles, California, Mike Davis (1990; 1996) documents the historical, architectural and mechanical transformation of an entire city, the human population and the natural environment. LA is divided and policed. LA has few public places, and the city's architecture assists in the surveillance of its inhabitants. When LA is flooded, its surrounding hills on fire, or when its inhabitants suffer another earthquake, the newspapers claim that "nature has let us down." To confront a chaotic and unpredictable landscape, unique features of the earth are paved over in accordance to the canons of bureaucratically organized natural resource management agencies and planning departments. Miles upon miles of shopping malls and mini-malls are linked in a series of superhighways and electronic corridors. From northern LA County to the Mexican border (an area of roughly 200 miles) everything looks the same and everything is in order. The natural landscape is buried and under concrete. The citified population is placeless. The mechanical landscape is without key reference points or geographical landmarks. Here, a tragedy of the senses unfolds – humanity is "unable to have direct contact with more satisfying means of living, tak[ing] life vicariously, as readers, spectators, passive observers" (Mumford 1938: 258).

Wilderness according to Gary Snyder (1990), is the place where the bears are. Those bears that walk LA's streets are considered far from their environment, are labeled "spoiled bears" and eventually removed. In place of the golden bear on the California flag, we construct Smoky-the-Bear, Teddy Bear and Pooh Bear. These signs of a bear are tamed while the city is denatured and dehumanized.

As a representation of the world-city, LA's reach is far beyond its economic and political boundaries, its hyperreality, or its Disneyfied urban–metropolitan design. As in all world-cities, the region's resources have been unsustainably exploited (Davis 1996; Sachs 1992). In order for the development to succeed over time, new markets (other regions' labor and resources) are imported and exploited (Daly and Cobb 1990; Shiva 1997). In return, the instrumental and mechanical values of globalism are exchanged and exported. The world-city depends on other markets that can be traded with and consumed on a multinational scale. More than resources are being traded in global economy. Economic, bureaucratic and technocratic ideas and values are also imposed on other cultures. Wild salmon or a free-flowing river matters less than domesticated pets and tropical "fish-by-mail." Multinational trade alliances, such as the North American Free Trade Agreement (NAFTA), support increasing levels of consumption/production in exchange for the values of modern technology, bureaucracy and economic rationality (Shiva 1997).

In global economy, diverse landscapes and cultural differences are exploited. William Knoke advocates globalization as an alternative to a sense of place and maintains that "we are entering the fourth dimension . . . an age of *Everything–Everywhere*. We are living in a placeless society . . . a superconnected society where distances cease to exist, where you can reach about and touch everyone in the world" (1996: 7). Knoke's message is one shared by the political and economic elite who glorify and perpetuate the myth of the benefits of the "global village." This global "village" is a highly centralized, bureaucratized economy that can be controlled by multinational corporations and states (Brecher and Costello 1994).

Global economy defines progress in term of the successful exploitation of new markets (in Indonesia, India, Mexico and Brazil), higher levels of consumption/production, and the new dependencies created in international alliances. International organizations such as the World Commission on Environment and Development (WCED), International Monetary Fund (IMF) and the World Bank propose "sustainable development" (SD). These organizations continue to prioritize "reviving economic growth" and ignore the "deeper socio-political changes (such as land use reform) or changes in cultural values (such as overconsumption in the North)" that would be required to sustain cultural and ecological diversity (Lelé 1991: 613). The terms of trade and order of the colonial era are maintained, and the impact on diversity is profound (Norgaard 1987). As a region's resources enter global markets, the level of resource use and rate of biological extinction accelerates. At a Honduras bus stop, indigenous people watch the soap operas "Santa Barbara" and "Dallas," and dream of capital goods and machinery. Placed on the hill top, the landmark in a local village reads "Coca Cola." Vandana Shiva (1997) argues that international trade agreements are forms of "biopiracy" that protect the northern hemisphere from the southern hemisphere so large-scale economic production/consumption can continue. The new world-city continues to grow, and its growth depends on other markets and other people. Entire bioregions – cultures and ecosystems – are developed to be exported and imported.

The globalism supported by the international trade alliances and the SD-promoting international monetary system is founded on a naive faith in science and an uncritical view of technology's ability to solve the human ecological crises (Goldsmith 1993). International organizations are promoting SD; at the same time, many of these globalists believe that modern society can technologically replace lost species and habitats. Gregory Stock's *Metaman: The Merging of Humans and Machines into a Global Superorganism* (1993) is a view of science and technology that can recreate and replace whatever "environment" global society wants – whether it be a pig with a human heart, a biosphere reserve, or a natural theme-park. The world is a machine, and metaman simply replaces the broken part.

Scientific and technological values are globalized when these forms of knowledge and craft, respectively, are separated from place or decontextualized. In a critical analysis of genetic engineering Craig Holdrege (1996) states, "We object-think when we focus on a detail or part of a larger system and then proceed to treat this part as an independent entity, even when we are trying to integrate it into a larger whole. The consequence is a mechanistic view of life and organisms." The decontextualizing characteristic of object-thinking is perpetuated by Stock, Knoke and other globalists. Globalists believe that technological and electronic society can create and model "ecosystems" on their computers while the earthly home is mechanically transformed to serve industrial and commercial ends.

But bioregionalists recognize that these technologies cannot free humanity from its dependence on the natural world, and cannot free humanity from the need to develop intimate relationships and partnerships with one another and nature. Stock and others fail to realize that metaman will not be able to technologically replace the human relationship to place if all that exists after economic development are inanimate machines, cyborgs, homogenized spaces and objects of human desire. The illusions of global economy encourage a parasitism that undermines the importance of the city, culture and the ecological community – in short, the bioregion – in human affairs. Lewis Mumford explains:

> Though man has become the dominant species in every region . . . partly because of the knowledge and the system of public controls over both man and nature he exercises, he has yet to safeguard that position by acknowledging his sustained and inescapable dependence upon all his biological partners.
>
> (1955)

A viable culture must find its roots somewhere, in *some place*. Bioregionalism and a sense of place lie beyond the faith in the perfect machine.

To "get our living together" (after Thoreau 1995: 69) within the context of globalism is no simple endeavor. There is the fear that given the power of globalism, bioregional values will be appropriated by the state. In *Sustainable America*, President Clinton's Commission on Sustainable Development describes the need for more decentralized, community-based environmental management. The Commission's proposal reflects a paradox: we attempt to expand our sense of community to include nature, but at the same time we are bound by the language, power and order of global economy. Metaman and the cyborg walk in a two-world structure of its own making; one world is governed in accordance with techno-bureaucratic and capitalistic values while the other world sways and evolves in accordance with the tug and pull of place. The individual within the industrialized context has grown dependent on various terms of order that structure personal and interpersonal relationships. These terms of order dictate that nature is to be organized, categorized and managed as if nature is a machine (or, as I have described earlier, denatured). The individual receives contradictory and incompatible information about the natural world. While nature is a machine to be controlled and managed, nature is also wondrous, wild, unmanageable, uncontrollable, chaotic, life-giving and life-producing. Although we attempt to know and value the bioregion differently, as bureaucratic consumers we cannot. Because of the dehumanizing character of globalism, the individual's inner psychological and spiritual sense of place and home become incoherent. Nature is bought and sold without a distinction made between one place or another. Our identity is no longer tied to place or our specific bioregion. Can we restore a sense of place to culture?

To restore the value of place in society, bioregionalists must reconcile a fundamental border redefinition conflict, which is depicted in Figure 4.3.

Spatial There are boundaries between the city and the countryside, between states, between the concepts of economy and the ecology, between private property, natural resources and wild nature, between predator and prey, between the human body and the mind, the past and the present, the present and the future, and the individual and society. Snyder writes: "The world of culture and nature, which is actual, is almost a shadow world now, and the insubstantial world of political jurisdictions and rarefied economies is what passes for reality" (1990: 37). Each boundary represents an alternative pattern of social and ecological negotiation.

Political boundaries may appear "hard" but they are only as "fast" as a watershed's boundaries. Salmon in the Columbia River will need to swim through seventeen distinct political jurisdictions to reach its spawning ground. To restore the salmon, a number of overlapping authorities and conflicting participants will need to cooperate. How do we reconcile these two different spatial scales? Do we continue to rely on top-down, highly centralized markets and bureaucracies or can we foster bioregionally-oriented relationships?

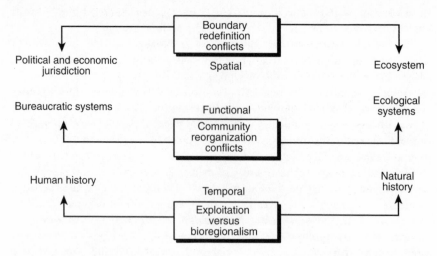

Figure 4.3 Boundary redefinition problems

Functional   Adaptation is a function of human and ecological communities (Rappaport 1977). This function is the same whether it occurs in organisms or societies. In societies under stress, conflict often precedes adaptation. Bureaucratic and capitalistic terms of order are coercively maintained and engineered, and they exhibit predictability. Natural systems function as adaptive, transitional systems which incorporate both predictable and unpredictable elements. A bioregion is hard to define, difficult to systematically understand and involves nonlinear processes that render attempts to quantify them difficult if not useless. How can we stop acting as functionaries of bureaucratically closed and maladaptive institutions and become functionaries of culturally and ecologically diverse bioregional systems?

Temporal   Modern institutions make decisions about the use of natural systems based on a limited time frame (e.g. political and economic cycles). Bioregional systems evolve in a much wider time frame. I would like to return to my earlier description of the problems associated with an instrumental value of a river. In a bioregionally-oriented culture, the river is more than a resource to be ordered for use. An ecologically literate community understands that a river is a source of life. Every living thing comes out of a watery environment. Water, as reflected in a free-flowing river, is bound by its edges, motions and patterns. In a watershed, we recognize that human life and sustenance depend on the health of a river. Rivers organically support distinct and interdependent communities that exist up and down their banks. Rivers are a source of cultural and ecological revitalization. These textual images hardly capture the feelings of flowing down a river, of being with a river, and of sharing in the ecology of "rivertime." These feelings and sensations of the river and watershed are essential features of bioregional living. We are part of the living systems we inhabit.

The wild salmon is an excellent sign of a healthy partnership between a culture and place. To integrate their community with place, several indigenous societies in the Pacific Northwest area of the US adopt the *totem* salmon. Totem salmon is celebrated in ritual, song and dance (House 1998 (forthcoming); Hay 1995). The salmon's return to spawn is a gift from nature to the tribe. The tribe's song and dance are gifts to the river, and for the generosity of the returning salmon. In dance, the salmon circle and cycle of life connect the past with the present and future.

Bioregions as self-organizing systems

In pursuit of bioregionalism, the real issue is not whether ecology can help bioregional theory-building and organization. It must. Ecological thinking can assist bioregionalists in developing sustainable cultural practices and organizations. Paul Shepard writes:

> Ecological thinking . . . requires a kind of vision across boundaries. The epidermis of the skin is ecologically like a pond surface or forest soil, not as a shell so much as a delicate interpenetration. It reveals the self ennobled and extended . . . as part of the landscape and the ecosystem.
>
> (1967: 2)

Ecological thinking not only demands interpenetration with community and nature but interpretation as well. Note that ecological terms like community, stability and hierarchy are imprecise and vague. There are at least four definitions of "species" – the biological, the evolutionary, the ecological and phylogenetic. Each have a different conceptual emphasis. Ecological thinking is as much a science as it is a craft and a sensibility because the ecology necessarily includes the intermingling of facts and values.

Social theorists have begun to stress the role of biological theory in reinventing institutions for the commons. In addition to condemning the bureaucratic leviathan, scholars (including Wilson 1975; Paehlke 1988; Rifkin 1991) have proposed alternative forms of organization. Lynton Caldwell (1987) offers the ideal of "biocracy" as an ecologically-oriented organizational form. Biocracy represents the application of biological principles to social organization. Caldwell contends that organizations should stress the values of diversity, stability, homeostasis and learning. Caldwell proposes that biologically-oriented social theories should represent the new organization paradigm, and that the various principles of biology are the key to understanding politics and society. The notion of "biocracy" represents the marriage of "bio" with "technocracy," and should be understood as a perpetuation of the cyborg identity which was described earlier. Caldwell's proposal for biocracy avoids the politics of knowledge and the power of knowledge. Who is going to interpret nature for society? Artists? Scientists? Poets? Nature should not be interpreted by those scientists and bureaucrats who lay claim to objectivity.

In *Rational Ecology*, John Dryzek is "interested in the capacity of human and natural systems *in combination* to cope with human-induced problems" (1987: 36; my emphasis). Human and natural systems in combination are bioregions. Like many bioregionalists, Dryzek is attempting to connect social and natural systems. Dryzek offers the criterion of "ecological rationality" to judge social systems (e.g. markets, administration). This criterion is based on the biological principles of negative feedback, coordination, robustness, flexibility and resilience. Dryzek is more interested in the "natural" and "social" dimension of systems than perpetuating the bureaucratic characteristics of organizing. Bioregionalists can learn from Dryzek's theory.

Building on the work of Dryzek among others, I propose the value of self-organization as an important characteristic of bioregionalism. A common characteristic of all life is a system's self-organizing capacity, or what biologists and system theorists refer to as *autopoiēsis*. Autopoiēsis is a term from the Greek words *auto* meaning "self" and *poiēsis* meaning to "make or produce." Autopoiesis is defined as the self-producing character of living systems (Zeleny and Hufford 1992). Self-organizing systems are dependent on the unity and relationships between the system's parts. This unity and relationship between parts is called autopoiesis. Notwithstanding their diversity, all living things are autopoietic insofar as all life continuously strives to regenerate its own organizational activity and structure. All life lives off other life. Warwick Fox describes the value of autopoiesis as follows: "[A]ll process-structures should be included in the class of living systems [which] open the door for the inclusion of ecosystems, species and social systems" (1990: 192). Together, all living processes coevolve and are interdependent. The focus here is on the living character of life as opposed to the mechanical sense endemic to the cyborg.

Poiēsis is manifested first of all in *phusis*. The Greek term *phusis* (which is the essential fullness of the natural world) is the "self-emerging and self-unfolding that lingers, endures, and prevails while simultaneously withdrawing into a self-closure that shelters and hence preserves the ongoing possibility of emerging" (Foltz 1995: 132). To put it another way, *phusis* is the "arising of something from out of itself" (Foltz 1995: 7). *Phusis* is a primary mode of *poiēsis* – the self-producing and self-withdrawing quality of life. A plant, for example, sprouts and emerges from the soil, and with the nutrients from the soil and sun extends itself into the open. This is the unfolding quality of life. In time, the plant will withdraw and return to the soil. The plant's organic qualities will become part of the soil, and will no longer be held within the boundaries of what we refer to as the "plant." In this sense, the plant goes-back-into-itself. This is the withdrawing characteristic of death. The plant and the soil are reunited. This is imperative for the reproduction of a healthy soil and plant life; life and death in an ecosystem coexist.

As in the case of all systems, human organizations should be adaptive, coevolving, complex and capable of creative action. Bioregional boundaries should reflect the self-producing and self-withdrawing characteristic of living systems. These values should be the criterion of a system's self-organizing character. The human pancreas, for example, is considered autopoietic insofar as it reproduces itself every twenty-four hours. The human body is a self-generating and self-renewing system of component parts that unite the mind, body and *life*. For both cultural and ecological (autopoietic) systems, the boundary is defined by the system's structures and processes, its very existence and its self-producing capacity. Unlike a bureaucratically and mechanically constructed boundary that divides the natural world into parts to manage, the boundary of a particular autopoietic system, like a cell, emerges as the system's components interact (Zeleny and Hufford 1992: 146).

With the value of self-organization, those human organizations which fail to interact, adapt, learn are maladaptive and are not autopoietic. Bureaucratically and mechanically oriented organizations may be self-perpetuating, but these forms of organizations are hardly autopoietic because when their boundaries dissolve, they do not reassemble themselves. Bureaucratic organizations are not systems for as Zeleny and Hufford show:

> It is both improper and unscientific to consider engineered social designs as *social* systems. Concentration camps, jails, command hierarchies, totalitarian orders, and so on, are not social orders but dictatorial, rule-based systems: everybody is put in place, told what to do and how to respond, where to go and when. Whatever social-system characteristics do emerge, do so only in spite and in defiance of the imposed order. There is nothing *spontaneously social* about them. As soon as the boundaries (the imposed rules, order, fear) are dissolved they do not reassemble themselves spontaneously: *rather everybody goes home*.
>
> (1992: 157; my emphasis)

Boundaries of self-organizing systems are not coercively maintained as they are in bureaucratic organizations (diZerega 1993). There is no threat of the use of power and punishment behind system reproduction. Because human beings and all life-forms belong to the class of autopoietic systems, the failure to sustain and preserve the autopoietic character of life represents a life-threatening act against the natural community. As important parts of a system of relationships, self-organizing systems are (1) far from thermodynamic equilibrium (e.g. a watershed's boundaries change over time); (2) governed by internal rules that support positive feedback, the exchange of energy and flexibility; (3) embedded in a network of larger-scale constraints (Hollick 1993).

Bioregionally-oriented forms of organizations should strive to organize on the basis of self-regulation and autopoiesis. This is not a romantic vision as some critics of bioregionalism propose (e.g. Dryzek 1997). Rather, the principle of autopoiesis fits well with bioregional science and sensibility. To sustain a social system, it should be the function of bioregionalists to enhance the capacity of the system for self-organization. To support autopoiesis, there must be unity and cooperation between individuals in a system. Varela, Maturana and Uribe show: "The establishment of the unity is logically and operationally *antecedent* to its reproduction" (1974: 189; my emphasis). The human body depends on the pancreas for survival. The pancreas depends on other organs to effectively process complex sugars. In ecosystems, predators depend on prey, creatures and systems depend on others for survival. In this sense, we are members of a "circle of animals."

Bioregional boundaries are constrained by the "reality" of the physical world that constantly changes in time, space and function. This view is hardly an ecologically deterministic one. Bioregions are constructs of a culture and community rather than biogeographical certainty. Bioregionalism grows out of the various perspectives and values held by the inhabitants of particular places. Although it may make sense for human beings to call themselves "Australian" or "Bostonian," other life-forms depend on a "greater place." Regional provincialism, akin to nationalism, should give way to the needs of others. Bioregionalists recognize that nonhuman beings have different perceptions of home place. Bioregional boundaries are defined only to be redefined in terms of changes in the character of the physiographic region, culture, history, current land-use pattern and climate (McCloskey 1989; McGinnis 1995b). Bioregionalism requires the *renaturalization* to foster the self-producing quality of human and natural systems. As autopoietic systems, bioregions are unbounded places "of the landscape" and culture. Bioregions are not static mechanical entities. A bioregion's boundary emerges when its inhabitants interact, react and process new information.

In accordance to the values of autopoiesis, bioregional organizations embrace three processes: (1) *production* (poiēsis): the rules and regulations guiding participation within a commons (such as membership, birth and acceptance); (2) *bonding* (linkage): the rules guiding participation, function and the positions individuals hold during their tenure within the organization; (3) *degradation* (disintegration): the rules and processes associated with termination of bioregional membership (death and separation). As ecosystems evolve, bioregional organizations and their membership shift into alternative social arrangements. A bioregional organization has a life cycle described as follows: the establishment of the organization, the maintenance of the organization, and the eventual decay of the organization. Bioregions move from a stable state through a zone of disruption to a relatively stable state (Schon 1971). Bioregional organizations operate under high entropy and accept the continued inputs from the community. As social systems, bioregional organizations must move in time and space, act and react.

A bioregional culture resists the coercive pressures and boundaries forced and maintained by mechanistic order. Such a resistance is possible. As in all shifts from one epoch, cultural myth, or dominant ideology to another, a choice between fundamental values is inevitable. During the industrial era, the further development of the mechanistic sense was one choice that had a significant impact on ecology and society. With new insights and values, the development of bioregional organization is another choice. Richard Tarnas describes the remarkable and creative evolution of Western world-views:

> The essential reality of nature is not separate, self-contained, and complete in itself, so that the human mind can examine it objectively and register it from without. Rather, nature's unfolding truth emerges only with the active participation of the human mind. Nature's reality is not merely phenomenal, nor is it independent and objective; rather, it is something that comes into being through the very act of human cognition.
>
> (Tarnas 1992: 434)

Nature's unfolding truth emerges in an identification with place, self-interest is reflected in the sustenance of place, the sensuousness and love of one's place, and in a rekindling of the childhood and primal memories of an earthly existence. The living sensuous world of place is always more direct, interactive and local than world-machines. We can resurrect a cultural relationship with the land, drawn from the senses, memories and moments each individual has with place. In the cognitive representations of place, we find the essential rudiments of bioregionalism – a sense of place and community reincarnate. To restore a sense of place, we must embrace home place – we must begin to reorganize.

Toward a restoration of the self in a bioregional organization

It is not easy to delete all the passages, stories and words of a place. The true test in reinhabiting place is in our personal and shared abilities to unwrap and tap the inner expressions, experiences and senses that collectively make up our cognitive maps of place. Cognitive maps are expressions of deeply held values, a culture's occupancy, and are creations of highly subjective processes. Cognitive maps are the medium between what is and what is becoming. The cognitive maps of the senses are hard to identify: they cross political, generational and ownership boundaries, and remain elusive. Nevertheless, these cognitive maps are the basis for a return to home place. Returning to home place requires a restoration of the self, a new-old mental continent.

Two major points are essential to restore a sense of self in a bioregional system:

Interior to exterior From the nucleus of the world-city, walls and waterways were built to separate civilization from the outside environment. The constructed boundaries protected humanity from the "wilds" of nature. Yet, these same boundaries are manifested as human behavior and institutional prac-tice (which, in industrial society, is reflected in language, management systems, law and our poor treatment of others). We speak of "nature" as an environ-ment, a natural resource, or wilderness because our contests with nature make us so afraid, in awe or fearful of its uprising. Nature remains an object to behold within the realm of aesthetics.

We are convinced that the world outside takes the same form as modern reason. Modern reason, which supports and justifies global civilization, remains on the surface of the world. One thinks the greatest value of mental function is to be found by dismantling and dissecting the world into all its parts, and studying separately each of these parts. Nature as an object of investigation remains far removed. The moment of human isolation from place occurs during usage. When nature-as-an-object is transformed into a dead remnant of human thought, nature becomes merely an instrument of elaboration.

We should recognize that there are many other ways of getting to know the world, and that the sum of the parts of a place does not equal the whole. When we wish to perceive of place really well, we need to regard it with its surroundings. The elements of place change position. We need to open up our perceptual field of vision to include the animate, breathing world. Bioregionalism requires the natural incorporation of interior with the exterior, and the field of bodily expansion to include others and place. Human activity takes shape with the winds, trees and rivers. We should recognize the connection between the interior (self) and the exterior parts of the landscape. The incorporation of the living breathing world with the human body requires a deeper form of human reason – a larger door to the "outside" – and understanding of our place in the world. As the painter Cézanne observed "nature is in the inside" or as Maurice Merleau-Ponty wrote "the world is made of the same stuff as the body" (1964: 163).

This marriage of the mind with ecology represents nothing less than a rewriting and reconstructing of the body and flesh. There is no human experience that exists outside of nature and culture. Bioregionalism is predicated on the movement of isolation to association, and the movement of the interior to the exterior.

Object-thinking to context-thinking   To acquire and sustain an open field for human activity, natural surroundings are part of our reference points, and exert influence on human activity. The human body is enclosed and wrapped in a bioregion. One cannot withdraw.

The recovery of a sense of place requires that society conjoin the study of the natural, social and personal components of human ecology, avoid one-dimensional fixations on "objective" matter, "subjective" mind, or the "intersubjective" construction of reality, and explore the world-making and self-making powers of language, perception and other vehicles of culture that can sustain place (such as dance, ritual, lovemaking, song, storytelling, sharing).

These are some pre-requisites for sloughing off your "European skin" to merge your human frame and mind with the landscape, and to once again feel the tug and pull of a fibrous and animate world. Only with the death of the machine, can the mind and body free itself to feel the tug and pull of place.

Despite the machines we have created and become, we must find a place to dwell.

> Life exists in the midst of a thousand transparent images – of white, blue tones folded over the universe, and dark shadows shaped by the wind and water. These caverns of the universe house millions of thoughts, memories and creature faces. These basins, drainages and canyons are mosaics that lead to an ocean of colors and clouds; and the blinking eyes of a thousand beings mirror the images of the self. These eyes reveal the memory of the land, shared laughter, our smallness and the glistening of a smile.
>
> . . . the days of sand wash over me. The sharp mountains smile with pine, the spring flowers dance with the colors of the day, and I smell the sage and ocean breeze. I feel at home, the wind blows the seeds of spring up to the mountains, whose jagged tops split the sky. My shadow bleeds into the soil.

Acknowledgments

I would like to thank Bronislaw Szerszynski, Tim Luke, Christina McGinnis, Beau Jack and two anonymous reviewers for their helpful comments on drafts of this essay.

Bibliography

Abram, D. (1996) *Smell of the Sensuous: Perception and Language in a More-than-Human World*, New York: Pantheon.

Artaud, A. (1993) "Fragments of a Journal in Hell," *Artaud Anthology*, San Francisco: City Lights.

Baudrillard, J. (1995) "The Virtual Illusion: Or the Automatic Writing of the World," *Theory, Culture and Society* 12: 97–107.

Bennett, J. (1993) "Primate Visions and Alter-Tales," in J. Bennett and W. Chaloupka (eds.) *In the Nature of Things: Language, Politics and the Environment*, Minneapolis: University of Minnesota Press.

Brecher, J. and Costello, T. (1994) *Global Village or Global Pillage*, Boston: South End Press.

Caldwell, L.K. (1987) *Biocracy: Public Policy and the Life Sciences*, Boulder: Westview Press.

Daly, H. and Cobb, J. (1990) *For the Common Good: Redirecting the Economy Towards Community, the Environment and a Sustainable Future*, London: Green Print.

Davis, M. (1990) *City of Quartz: Excavating the Future in Los Angeles*, London and New York: Verso.

—— (1996) "How Eden Lost Its Garden: A Political History of the LA Landscape," *Capitalism Nature Socialism* 6 (4): 1–30.

diZerega, G. (1993) "Unexpected Harmonies: Self-Organization in Liberal Modernity and Ecology," *Trumpeter* 10 (1): 25–32.

Dodge, J. (1981) "Living By Life: Some Bioregional Theory and Practice," *CoEvolution Quarterly* 32: 10–2.

Dryzek, J.S. (1987) *Rational Ecology: Environment and Political Economy*, New York: Basil Blackwell.

—— (1997) *The Politics of the Earth: Environmental Discourses*, New York and Oxford: Oxford University Press.

Durning, A. (1992) "Guardians of the Land: Indigenous Peoples and the Health of the Earth," *Worldwatch Paper 112*, Washington, D.C.

Eco, U. (1994) *How to Travel with a Salmon and Other Essays*, New York: Harcourt Brace.

Espeland, W. (1994) "Legally Mediated Identity: The National Environmental Policy Act and the Bureaucratic Construction of Interests," *Law & Society* 28 (5): 1149–79.

Foltz, B.V. (1995) *Inhabiting the Earth: Heidegger, Environmental Ethics and the Metaphysics of Nature*, New Jersey: Humanities Press.

Fox, W. (1990) *Towards a Transpersonal Ecology*, Boston: Shambhala.

Goldsmith, E. (1993) *The Way: An Ecological World-view*, Boston: Shambhala.

Gomez-Pompa, A. and Kraus, A. (1992) "Taming the Wilderness Myth," *Bioscience* 42 (4): 271–9.

Harraway, D.J. (1985) "A Manifesto for Cyborgs: Science, Technology and Socialist Feminism in the 1980s," *Socialist Review* 15 (2): 64–107.

Hay, J. (1995) *A Beginner's Faith in Things Unseen*, Boston: Beacon Press.

Heidegger, M. (1977) *The Question Concerning Technology and Other Essays*, trans. W. Lovitt, New York: Harper & Row.

—— (1980) *Holderlins Hymnen "Germanien" und "Der Rhein" (1934–1935)*, in S. Ziegler (ed.) *Gesamtausgabe*, part II, vol. 39, Frankfurt: Klostermann.

Holdrege, C. (1996). *Genetics and the Manipulation of Life: The Forgotten Factor of Context*, Hudson: Lindesfarne House.

Hollick, M. (1993) "Self Organization and Environmental Management," *Environmental Management* 17 (5): 621–8.

House, F. (1998, forthcoming) *Totem Salmon*, Boston: Beacon Press.

Hummel, R. (1994) *The Bureaucratic Experience*, New York: St. Martin's Press.

Kellert, S.R. (1996) *The Value of Life: Biological Diversity and Human Existence*, Washington, D.C.: Island Press/Shearwater Books.

Kemmis, D. (1990) *Community and the Politics of Place*, Norman: University of Oklahoma Press.

Kimberly, J.R., Miles, R.H. and associates (1980) *The Organizational Life Cycle*, San Francisco: Jossey–Bass.

Knoke, W. (1996) *Bold New World: The Essential Road Map to the 21st Century*, New York: Kodansha International.

Lelé, S.M. (1991) "Sustainable Development: A Critical Review," *World Development* 19 (6): 607–21.

Lopez, B.H. (1978) *Of Wolves and Men*, New York: Charles Scribner's Sons.

Ludwig, D., Hilborn, R. and Walters, C. (1993) "Uncertainty, Resource Exploitation and Conservation: Lessons from History," *Science* 260: 17–9.

Lukas, D. (1996) "A Place Worth Caring For," *Orion Society Notebook*, Autumn/Winter: 18.

Luke, T.W. (1996) "Liberal Society and Cyborg Subjectivity: The Politics of Environments, Bodies and Nature," *Alternatives* 21: 1–30.

Mander, J. and Goldsmith, E. (ed.) (1996) *The Case Against Global Economy: and for a Turn Toward the Local*, San Francisco: Sierra Club Books.

McCloskey, D. (1989) "On Ecoregional Boundaries," *Trumpeter* 6 (4): 127–32.

McGinnis, M.V. (1994) "Myth, Nature and the Bureaucratic Experience," *Environmental Ethics* 16: 425–36.

—— (1995a) "On the Verge of Collapse: Wild Salmon, the Columbia River System and the Northwest Power Planning Council," *Natural Resources Journal* 35 (1): 520–52.

—— (1995b) "Bioregional Organization: A Constitution of Home Place," *Human Ecology Review* 2 (1): 72–84.

Merleau-Ponty, M. (1964) "Eye and Mind," ed. and intro. J.M. Edie, *The Primacy of Perception*, Evanston: Northwestern University Press.

Mills, S. (1995) *In Service of the Wild*, Boston: Beacon Press.

Mumford, L. (1938) *The Culture of Cities*, New York: Harcourt, Brace.

—— (1955) "The Natural History of Urbanization," ed. W.L. Thomas with C.O. Sauer, M. Bates and L. Mumford, *Man's Role in Changing the Face of the Earth*, proceedings of international symposium, Chicago: University of Chicago Press.

Norgaard, R.B. (1987) "Economics as Mechanics and the Demise of Biological Diversity," *Ecological Modeling* 38: 107–21.

Noss, R.F. *et al.* (1995)*Endangered Ecosystems of the United States: A Preliminary Assessment of Loss and Degradation*, Biological Report 28, Washington, D.C.: US Department of the Interior.

Odum, E. (1971) *Fundamentals of Ecology*, Philadelphia: Saunders.

Paehlke, R. (1988) "Democracy, Bureaucracy and Environmentalism," *Environmental Ethics* 10: 304.

Palmer, T. (1993) *Lifelines*, Washington, D.C.: Island Press.

Poster, M. (1995) *The Second Media Age*, United Kingdom: Polity Press.

Rappaport, R. (1977) "Maladaptation in Social Systems," in J. Friedman and M.J. Rowlands (eds.) *Evolution of Social Systems*, London: Duckworth.

Rifkin, J. (1991) *Biosphere Politics: A New Consciousness for a New Century*, New York: Crown.

Sachs, W. (ed.) (1992) *The Development Dictionary: A Guide to Knowledge as Power*, London: Zed Books.

Schon, D.A. (1971) *Beyond the Stable State*, New York: W.W. Norton.

—— (1967) "Ecology and Man: A Viewpoint," in P. Shepard and D. McKinley (eds.) *The Subversive Science: Essays Toward an Ecology of Man*, New York: Houghton Mifflin.

Shepard, P. (1996) *The Others: How Animals Made Us Human*, Washingon, D.C.: Island Press.

Shiva, V. (1997) *Biopiracy: The Plunder of Nature and Knowledge*, Boston: South End Press.

Snyder, G. (1990) *The Practice of the Wild*, San Francisco: North Point Press.

—— (1995) *A Place in Space: Ethics, Aesthetics and Watersheds*, Washington, D.C.: Counterpoint.

Stock, G. (1993) *Metaman: The Merging of Humans and Machines into a Global Superorganism*, New York: Simon & Schuster.

Tarnas, R. (1992) *The Passion of the Western Mind: Understanding Ideas That Have Shaped Our World View*, New York: Ballantine Books.

Taussig, M.T. (1987) *Shamanism, Colonialism and the Wild Man: A Study of Terror and Healing*, Chicago: University of Chicago Press.

Thoreau, H.D. (1995) *Walden*, Boston and New York: Houghton Mifflin.

Varela, F.J., Maturana, H.R. and Uribe, R. (1974) "Autopoiesis: The Organization of Living Systems, Its Characterization and a Model," *Biosystems* 5: 187–96.

Weber, M. (1958) *The Protestant Ethic and the Spirit of Capitalism*, trans. T. Parson, New York: Charles Scribner's Sons.

—— (1973) *Essays in Sociology*, trans. H.H. Gerth and C. Wright Mills, New York: Oxford University Press.

White, R. (1995) *The Organic Machines*, New York: Hill & Wang.

Wilson, E.O. (1975) *Sociobiology: The New Synthesis*, Cambridge: Belknap Press of Harvard University Press.

Zeleny, M. and Hufford, K.D. (1992) "The Application of Autopoiesis in Systems Analysis: Are Autopoietic Systems also Social Systems?" *International Journal of General Systems* 21: 145–60.

5 Bioregional possibilities in Vermont

Christopher McGrory Klyza

One of the major problems with theories calling for significant changes in the way modern societies and institutions are designed is that they are too abstract, removed from practical concerns and issues. This is true of bioregionalism. In order to take this approach to the next level, we need to show how bioregionalism might work on the ground, in specific places. Vermont – a place where natural and human communities are fairly healthy – is an interesting place to explore the potential for translating bioregional theory into practice. It is the most rural state in the US; it has the second smallest population of any state; it has a tradition of vibrant local democracy, most apparent in the continuing town meeting; it has a strong independent streak (Vermont was an independent republic from 1777 to 1791); and its landscape is one that is recovering from massive ecological changes wrought by European settlement.

A key to making bioregionalism work is a close examination of boundaries and what they mean. As Mike McGinnis argues in Chapter 4, all boundaries are human constructs. I don't think we are particularly well served if we simply use bioregionalism to substitute one more ecologically rational set of boundaries for an existing set of ecologically less relevant boundaries. Rather than wed ourselves to new boundaries, I hope that bioregionalism can help us focus on the fundamentally social nature of boundaries and to think of overlays of borders and boundaries.

In this chapter, I will:

1   examine the current status of the natural world in Vermont and the surrounding region, since one of the key components to a bioregional approach is the integration of healthy human communities with healthy natural communities;
2   discuss a number of significant bioregional initiatives in the region, ranging from the 26-million-acre Northern Forest Lands Study to the formation of local watershed groups;
3   illustrate boundary overlays by examining the bioregional layers in Vermont, suggesting which borders might be most useful for particular purposes; and

4   offer concluding thoughts on the prospects for bioregionalism in Vermont and discuss how this case can help us to think about bioregionalism in other regions of the world.

The natural scene

One of the key reasons – if not *the* key reason – for moving in a bioregional direction is to improve the interaction between humans and nature, to strive for a place in which natural and human communities are sustainable. On this point, Vermont and northern New England and New York offer an extremely interesting testing ground for bioregionalism. The forests in this region have returned and, in many places, species long absent have returned as well. As Bill McKibben writes: "If you're looking for hope, this unintentional and mostly unnoticed renewal represents the great environmental story of the US – in some ways, of the whole world" (1995: 13). The land is recovering its integrity, and a "rewilding" is occurring in many places (Klyza 1994a).

Looking south down the spine of the Green Mountains from Mt. Abraham, which rises above my home in Bristol, my view is of green velvet. Roughly 80 percent of Vermont is now covered by forest, which is down from the 95 percent when the Europeans arrived, but up greatly from the nadir of approximately 35 percent forest cover around 1870. A similar pattern is found in New Hampshire (87 percent forested, up from a low of 45 percent) and Maine (89 percent forested, up from a low of 77 percent). In McKibben's words again: "The ghost map of this place is reasserting itself" (1995: 16). In Vermont, five animals are known to have been extirpated since the European arrival: Arctic char, caribou, elk, timber wolf and wolverine. A sixth, the passenger pigeon, is extinct as well (Trombulak 1995). In the rewilding of this region, several species extirpated from the state – beavers, fishers, moose, ospreys, peregrine falcons, wild turkeys and others – are returning, some with human help, some without.

State and federal agencies are involved in some of these restoration projects, such as reintroducing peregrine falcons and wild turkeys. Resource agencies have also played a major role in restoring and protecting habitat in state forests and the Green Mountain National Forest, with its collection of small Wilderness Areas. Private groups, like the Green Mountain Forest Watch, the Nature Conservancy and the Vermont Natural Resources Council play major roles as well. Just as important, though, has been the unmanaged and unplanned rewilding of much of the state as abandoned farms and pastures return to forest. It has been this process that has drawn the moose back into Vermont in large numbers, moving down the spine of the Green Mountains to disperse throughout the state. Beavers, nearly extirpated by trapping and habitat destruction, are abundant again thanks in large part to natural rewilding.

However, this process of rewilding takes time. The healthy forest of the past is not the forest of today. The forest of today's Vermont and northern New England is a different forest in terms of structure and size distribution; it is denser than the forest of the 1600s, composed of smaller and younger trees. It might be considered the "second" forest of this region. The forest is at an earlier stage of ecological succession: only 1,500 acres of old growth remains in Vermont. The composition of the trees in the forest has significantly changed, with chestnut and elm virtually gone and introduced diseases and insects taking a significant toll on many other species (e.g. butternuts). There has been a decrease in beech and red spruce, an increase in aspen, red and sugar maple. Forest fragmentation reduces the size of undisturbed habitat with significant effects on a variety of species (such as many migrating forest songbirds). Eighty-eight plant species are known only from historical records, and 13 percent of conifers, 43 percent of ferns and allies, and 36 percent of flowering plants are classified as rare, threatened, or endangered. Within the state 17 percent of birds, 34 percent of fish, 28 percent of mammals, and 46 percent of reptiles and amphibians are classified as endangered or threatened with extinction. One in four vascular plants in Vermont is an exotic (or non-native), and many exotics cause significant harm to native trees (such as the gypsy moth and the hemlock woolly adelgid) as well as leading to the displacement of native species (e.g. the European buckthorn). Finally, this returning forest is under stress from exploitive logging in some places, development in other places, air pollution including acid rain throughout, and perhaps the beginnings of global climate change (Trombulak 1994; 1995).

Nonetheless, Vermont and the northern New England–New York region offer a chance to "connect with land in recovery, develop an intimacy with land that we have severely damaged in the past" (Klyza 1994a: 78). McKibben writes:

> Nature's grace in the American East offers this hope to a world in terrible need of models. For the East is a typical place – a place where large populations live in and around the recovering woods and rivers. In that way, it looks like the rest of the world. . . . In place of the increasingly sterile debate between wilderness and civilization, between raped and virgin, it offers at least the outside possibility of marriage.
>
> (1995: 15)

This area can serve as a focus for our thinking and work about the role for humans *with* nature; it can help us respond to William Cronon's criticism (1996) that modern environmentalism is too focused on a culturally-constructed wilderness without humans. Here humans do – and can – live amongst the wild, part of a common fabric. As discussed above, all is not well in the forests of this region, but we have a fragile chance to regain and restore what has been lost.

A bioregional approach, I think, offers us the best chance to make good this fragile chance. At its core, bioregionalism integrates the relationship of culture and the natural world. Bioregionalists are attuned to their place in the landscape, and more likely to care about the land and commons. They are more likely to put in the hard work to restore and protect forest habitats – including more wild areas, so the catamount and wolf can one day return – *and* to recreate our economies and communities, so that perhaps we can make the transition to sustainable human and natural communities.[1]

Recent bioregional initiatives

With these changes in the natural scene as both backdrop and catalyst, there have been a number of forays into regional and bioregional thinking and action in Vermont and northern New England in recent years. These different approaches can be thought of as bioregional layers, based on different scales and different cultural and natural borders. At the largest scale, the Northern Forest initiative, the Lake Champlain Basin Program, and the Champlain–Adirondack Biosphere Reserve have helped people to see beyond political borders and to make more connections based on natural landscapes. There has also been much action at the local level, with groups forming at small watershed levels.

The Northern Forest as a regional (though not a bioregional) entity was born in the fall of 1988, when the US Department of Agriculture Forest Service and the Governors' Task Force on Northern Forest Lands – charged by Congress and the four governors – each began studying land ownership, conservation strategies, forest resources, economics and the human communities in the 26-million-acre region spreading across four states. The catalyst for its creation was a series of corporate takeovers of forest products companies and subsequent land sales in Maine and the northern portions of New Hampshire, Vermont and New York. This set off fears throughout the region of land conversion – changing land from working forests to second home subdevelopments (Klyza 1994b). This is an area in which the forest products industry played a significant role in the economy and in which residents – and visitors – had been used to having free access to these largely private forests for recreation. The boundary for this Northern Forest was defined to focus on "the parts of the four states where large forested tracts (greater than 5,000 acres) were predominant," incorporating entire counties for ease of data collection (Harper, Falk and Rankin 1990: 91). This region is quite different from most other large forested regions in the US due to the very limited federal ownership – only a little over 1 percent. About 16 percent of the land is in public ownership, the great bulk of that (11 percent) in owned by New York State in the Adirondack Park. Over 50 percent of the land is owned by industrial and large nonindustrial owners. In Vermont, the Northern Forest included 2 million acres in five northern counties. Of course, the boundaries selected for the Northern Forest Lands Study are completely arbitrary from an biogeographical perspective. A similar forest spills into the Canadian provinces of Ontario, Quebec and New Brunswick, and encompasses virtually all of Maine, New Hampshire and Vermont, as well as the Berkshires in Massachusetts and the Catskills in New York. Nevertheless, the Northern Forest initiative presented a great opportunity for people in northern New England to start thinking about connections to the landscape that did not stop at state borders.

In the spring of 1990, the Northern Forest Lands Study and the Governors' Task Force each presented its findings. Combined, these reports recommended the creation of a Northern Forest Lands Council (NFLC) for four years to conduct further studies of the region. This council was created and it worked from the fall of 1990 through the fall of 1994, when it issued its findings and disbanded (Northern Forest Lands Council 1994). Throughout this four-year period, citizens from the area were actively involved in advising the NFLC. The council created four state Citizen Advisory Committees and held several public hearings and listening sessions. In addition, regional and national environmental groups formed the Northern Forest Alliance (with an ever-changing thirty or so groups) in 1990 to coordinate and focus attention on the Northern Forest. During this four-year discussion, the original focus on forest conversion was expanded to a more widespread discussion of biological diversity, forest health and forest practices. Many citizens in the region were disappointed that the NFLC disbanded, hoping that the council would become a more permanent forum to further underscore the "regional identity" of the forest. This option, however, was never seriously considered, in part due to strong opposition from property rights advocates from throughout the region. Instead, the NFLC recommended the creation of state forest roundtables as a vehicle to carry on its work and to implement its recommendations. From a bioregional perspective, this is most unfortunate since state borders, which had briefly appeared somewhat porous, were once again made the primary constraint for thinking about human interactions with the landscape.[2]

The Lake Champlain Basin Program (LCBP) grew out of a federal law passed in 1990 that was designed "to develop a comprehensive pollution prevention, control, and restoration plan for Lake Champlain" (Lake Champlain Basin Program 1994: Appendix A-1). The basin, roughly 5.3 million acres, is located in Vermont (56 percent), New York (37 percent) and Quebec (7 percent). The basin population of over 600,000 people is approximately two-thirds rural (based on the US Census definition: people living in towns with fewer than 2,500 people). The LCBP is guided by the Management Committee, thirty-one people representing the two states and relevant federal agencies. There are also state Citizen Advisory Committees and a Lake Champlain Steering Committee, an entity created in 1988 when New York, Quebec and Vermont signed a Memorandum of Understanding on Environmental Cooperation on Lake Champlain. This complex intergovernmental structure must deal with over 200 local and regional agencies with jurisdiction over land and water issues in the basin, as well as numerous state, provincial and national agencies.

The LCBP has focused primarily on water-quality issues in the lake, identifying three main concerns: high phosphorus levels leading to algae blooms, toxic pollutants such as PCBs and mercury, and nuisance non-native aquatic plants (e.g. water chestnut) and animals (e.g. zebra mussels). In developing a plan to address these and other concerns, the LCBP has, among other themes, relied on a watershed-based ecosystem approach. Particularly with the phosphorus problem, these approaches are crucial since most of the phosphorus entering the lake is due to nonpoint source pollution, which means land-use practices within the watershed must be altered in order to improve the water quality. As the Lake Champlain Basin Program explains:

> Action based on watershed boundaries rather than political boundaries, such as town or county borders, can better target polluted or threatened areas for restoration or protection. Citizens can then act to improve water quality based on their knowledge of their local area, and neighboring communities can link together to develop innovative ways to solve pollution problems within their watershed.
>
> (LCBP 1994: Introduction 3)

To help achieve this, the LCBP plan calls for building the capabilities for local watershed planning and protection. In addition, seven major sub-basins to the Lake Champlain Basin are identified that could prove to be useful as second-tier bioregions in those parts of Vermont and New York in the Lake Champlain Basin.[3] The focus of the LCBP on watersheds, ecosystems, combining ecological and economic components, and encouraging public involvement make this program a very valuable one in terms of getting people in the area to think and to act bioregionally.

After ten years of planning and discussion by citizens from the region, the Champlain–Adirondack Biosphere Reserve was designated in 1989 as part of the UN Man and Biosphere Program. The 10-million-acre reserve includes all of the Adirondacks and all of the Champlain Basin, except for the portion in Quebec. As of 1991, it was the fourth largest and most populated reserve in the world. These reserves focus on conservation, research and monitoring, and are important proving grounds for sustainable development. Little has happened in the reserve due to its designation. Since designation carries no funding with it, people in the region have focused on existing entities (e.g. the Adirondack Park) and new initiatives (e.g. the Northern Forest Lands Study/NFLC and the LCBP).

The Adirondack Park, across Lake Champlain from Vermont, offers an interesting case of a bioregional approach to integrating human habitation with wild lands. The park, established in 1885, features a mix of roughly 2.6 million acres of state-owned land and 3.4 million acres of private land (nearly 80 percent of which is open land). The state lands – the Adirondack Forest Preserve – are declared "forever wild" in the New York State constitution and managed as wilderness or wild forest. Over 135,000 people live amongst this scattered wild forest on a year-round basis, with 70,000 seasonal dwellers and millions of tourists each year (Collins 1994). The Adirondacks are "the world's first experiment in restoring an entire ecosystem" (McKibben 1995: 30). Although there have been difficulties during the last twenty years over restrictions on private land-use and economic development more generally, the Adirondack Park offers one of the best examples in the world of integrating humans into a wild ecosystem and serves as an important bioregional model.[4]

There have been bioregional initiatives from the bottom-up in Vermont as well. For instance, the Lewis Creek Association, founded as the Lewis Creek Conservation Commission – a coalition of five town conservation commissions – in 1990, focuses on the creek and its watershed as a cultural and natural resource.[5] Its goals are to promote improved water quality and protect fish and wildlife habitat, while encouraging sustainable land-uses. With a steering committee of twelve and a mailing list of over 500 households, the group has succeeded in focusing attention on the creek and its watershed as entities that extend beyond political boundaries (Henzel 1996). Another example of a group focused on a smaller watershed is the Friends of the Mad River. Serving a water-shed of 5,800 people and over 90,000 acres, the group recently joined with the Mad River Valley Planning District to complete a comprehensive conservation plan to protect and restore the Mad River watershed (Mad River Valley Planning District and Friends of the Mad River 1995). A bioregional organiza-tion with a larger focus is the Watershed Center. Founded in 1995, this group is committed to working for sustainable human and natural communities in the Lewis Creek, Little Otter Creek and New Haven River watersheds. Among the projects this group is pursuing are purchasing a piece of land to serve as an envi-ronmental education center for local schools; sponsoring the new Vermont Family Forests initiative, which is designed to encourage sustainable forestry, especially among small, private landowners; and helping to test a new biological integrity index for forest ecosystems developed by Steve Trombulak that would be used to approximate forest health throughout the region.

As the various groups, projects and studies discussed in this section suggest, there are numerous bioregional initiatives underway in Vermont and the surrounding area. These bioregional approaches reflect a kind of bioregional layering: from small watersheds (e.g. Lewis Creek Association), to medium-sized groupings of watersheds (e.g. Watershed Center), to drainage basin initiatives (e.g. Lake Champlain Basin Program), to partial ecoregion initiatives (e.g. Northern Forest Lands Study). We need to be thinking bioregionally at each of these levels, for it is only when people start to think bioregionally that we can start to reorient our institutions in a bioregional direction.

Uncovering Vermont's bioregional "layers"

In light of the previous discussions about the recovery of the natural landscape and the series of significant bioregional initiatives in Vermont, it is time we turned to envisioning a bioregional future in Vermont. When thinking about bioregionalism in Vermont, the first thing one must do is to figure out the scale of the bioregion and its natural and cultural boundaries. The scale in Vermont is much different from California, for instance. In his splendid essay "Coming in to the Watershed" (1992), Gary Snyder talks of six bioregions in California, suggesting that large parts of the state are better considered as part of other bioregions (such as the Great Basin and the Lower Colorado drainage). Each of these six California bioregions is based on rather distinct natural communities, leading "to different sorts of rural economies"; each bioregion encompasses millions of acres. The entire state of Vermont, roughly 6 million acres, is smaller than a number of these California bioregions. Snyder offers a keen cultural reference for bioregional borders, noting that "types of hats or rain gear go by the watershed" (1992: 69). Again, such a cultural barometer for bioregion borders makes little sense in Vermont, and indeed in much of the northeast, where precipitation is abundant and follows roughly similar patterns. What this brief comparison indicates is that California and Vermont are very different places, in terms of both natural and cultural communities, and that the bioregional borders in each place must reflect these differences.

For the remainder of this section, imagine yourself sitting at a geographic information system (GIS) terminal. With different commands, you can pull up different maps of Vermont and surrounding areas. Depending on the characteristic you have selected (e.g. home territory of indigenous peoples or air pollution levels), different borders on the landscape will make sense. None of these borders is inherently correct; the border that makes the most sense depends on the issue you are addressing. In the next few pages I will discuss the most significant of these borders from a bioregional perspective. Then, I will combine these different overlays to develop a system of bioregional "layering" that makes sense for Vermont.

Let's begin with potential natural borders in Vermont, based on geology, ecoregions and watersheds. Geologically, Vermont is quite complex for such a small place (Trombulak and Klyza 1998: ch.1). It is home to parts of three mountain ranges that it shares with New York, Massachusetts, Connecticut, New Hampshire and Quebec. The main range and the dominant geographic feature of the state is the Green Mountains. This range runs north–south throughout the center of the state, extending – geologically – into Massachusetts and Quebec. With its related parallel ranges, the Greens are from twenty to thirty-five miles wide. The Taconic Mountains begin in west central Vermont and extend into New York, Massachusetts and Connecticut. And in northeastern Vermont, highlands are created where the White Mountains complex spills over from New Hampshire. All of these ranges have different geologic and bioregional histories. For instance, the Green Mountains help define the pastoral image central to Vermont's tourist appeal and the Taconics are the center of the state's marble industry. The geologic break is even greater with neighboring New York. Lake Champlain, which forms over 100 miles of the Vermont–New York border, sits along thrust faults, indicative of a stark geologic boundary. The Adirondack Mountains are wholly unrelated to those in Vermont. They are younger mountains of very old rock, related to the Canadian Shield to the north rather than to the Appalachian chain to the east and south. So, in northern New England at least, the boundary between New York and New England makes geologic sense.

Which ecoregion Vermont is considered to be part of depends on the mapping scheme, its underlying principles and its scale. In Robert Bailey's first "Ecoregions of North America" (Bailey and Cushwa 1981), all of Vermont – as well as almost all of northern New England, most of New York, and parts of the Maritime provinces, Quebec, Ontario, Pennsylvania, Michigan, Wisconsin and Minnesota – is in the Laurentian Mixed Forest Province. The ecoregions on this map, according to Bailey, "correspond to broad vegetation regions having a uniform regional climate and the same type or types of zonal soils." Bailey's revised version of US ecoregions refines this ecoregion, retaining the Laurentian Mixed Forest Province for the New England lowlands – including the Champlain Valley in Vermont – and parts of New York, Pennsylvania, Michigan, Wisconsin and Minnesota. He adds the Adirondack–New England Mixed Forest–Coniferous Forest–Alpine Meadow Province, which covers the mountain regions of the original ecoregion – the Adirondacks and Catskills, Berkshires, Taconics, Whites, Maine Uplands and the Green Mountains and almost all of the rest of Vermont except the Champlain Valley; and the Eastern Broadleaf Forest Province, which includes a small portion of west central Vermont. Bailey has added cultural ecology, disturbance regimes, fauna, land-surface form, land-use and surface water characteristics to his original factors to help him refine this map (Bailey 1995). These provinces have been further refined by others to the section and subsection levels. In Vermont, there is one section (divided into two subsections) of Laurentian Mixed Forest Province, three sections (with eight subsections) of the New England–Adirondack Province, and one section (with two subsections) of the Eastern Broadleaf Forest Province (Smith *et al.* 1995). In James Omernik's ecoregion scheme (1987), based on land-use, land surface form, potential natural vegetation and soils, most of Vermont is part of the Northeastern Highlands, with the Champlain Valley designated part of the Northern Appalachian Plateau and Uplands and the Connecticut Valley halfway up the Vermont–New Hampshire border falls into the Northeastern Coastal Zone ecoregion. These ecoregion approaches are cast at *too large a scale* to adequately represent the different ecological communities in the state. Vermont is primarily hemlock–northern hardwoods (beech, birch and maple), spruce-fir forest in the northeast and mountain areas (above 2,500 feet), and smaller wetland communities (Trombulak and Klyza 1998: ch.2).[6] This ecoregion mapping is very valuable, but it should be viewed as an important piece in the bioregional puzzle and not the answer to it since the ecoregional sections, for instance, cut across watershed and cultural boundaries.

It is the watershed that people most often turn to when they are thinking of a bioregion's boundaries. Vermont is part of three main watersheds, which, like its mountains, are shared with its political neighbors (Meeks 1986). The St. Lawrence River, running to the north of Vermont through Quebec to the Gulf of St. Lawrence, drains 55 percent of the state. In Vermont, most of this drainage first runs into Lake Champlain, which drains north through the Richelieu River into the St. Lawrence. This is referred to as the Champlain

Basin (which also drains a significant amount of New York and a small portion of Quebec). A small part of northern Vermont also drains more directly into the St. Lawrence through Lake Memphremagog. The Connecticut River, flowing south into Long Island Sound, drains the eastern portion of the state (41 percent), as well as much of New Hampshire, Massachusetts and Connecticut. Finally, a small portion of the southwestern corner of Vermont (4 percent) is part of the Hudson River watershed, which empties its waters into the Atlantic.

Before further refining the proper scale and border of Vermont bioregions, we need to examine some of the cultural overlays developed by humans living there.[7] The territory of the Native Americans in the Vermont region at the time of European arrival (ca. 1600) offers some interesting insights into bioregional borders. The Western Abenaki lived throughout most of the state; the exception being the southwestern corner – the area that is part of the Hudson watershed – which was occupied by the Mahicans (Haviland and Power 1994). The Green Mountains served as a natural border as well, with different Abenaki bands located in the Champlain Basin and the Connecticut Valley. Lake Champlain was an even more significant border: to its west were the Iroquois, the traditional enemy of the Abenaki, while to the east were the Abenaki and the larger grouping of Wabanaki people throughout northern New England and the Maritime provinces. Hence, the lake presented a significant cultural border, one that remains in the distinction between New England and New York.[8]

Unlike the original thirteen states, Vermont had no royal colonial charter from England. This absence led to great controversy over the political future of Vermont until it became the fourteenth state in 1791. Snyder writes that: "The political boundaries of the Western states were established in haste and ignorance" (1992: 67). Yet, it was not just the western states, but all of the states (with perhaps the unconscious exception of Hawaii) whose boundaries were determined without real respect for natural landscapes. Indeed, the boundaries of the original colonies were drawn in London, often with an even fainter knowledge of the land than possessed by those in Washington, D.C., who drew the boundaries of California or Montana. There was also great confusion and controversy over the specific boundaries of the colonies in America. Massachusetts and New Hampshire claimed parts of Vermont, while New York claimed that, based on its royal charter, all of Vermont was actually part of New York (Trombulak and Klyza 1998: ch.3). The king ruled in favor of New York, dismissing the claims of the other two states. The settlers in Vermont did not agree with this decision, however, and in 1777 they declared themselves an independent republic, settling on the name Vermont after using New Connecticut for a few months. During its period of independence, Vermont even expanded three times – twice to the east, once to the west. So, the borders of the place we call Vermont were much disputed and not firmly settled until a little over 200 years ago. During these disputes, two interesting points are apparent. First, the distinction between Vermont and New York, this time culturally and politically, was again an important one. Second, the expansions to the east and another effort to create a new state centered on the Connecticut Valley demonstrated a respect for the landscape in the 1700s in the region. Those living on either side of the river recognized that a political entity based on the Upper Connecticut watershed made sense since the Green Mountains cut the region off from the western portion of Vermont and those living on the east side of the river felt politically isolated given the rest of New Hampshire's focus on the seacoast.

Vermont grew quickly in its early years as a state, but this growth had slowed dramatically by the mid-1800s. In 1830, over 280,000 people lived in Vermont, triple its population at statehood in 1791. In 1900, the population was a little more than 330,000, and from 1910 to 1920 and 1930 to 1940, the state's population fell. (Throughout this period, the population of the US was growing dramatically.) It was only around 1960 that Vermont started to undergo significant growth again, increasing in population from 390,000 to 580,000 in 1994 (Trombulak and Klyza 1998: chaps.3–6). This population has always been a rural one. Even today, Vermont is comfortably the most rural state in the country (67.8 percent). Its northern New England neighbors, Maine and New Hampshire, are third and seventh most rural (Morgan, Morgan and Quitno 1994: 414). In addition, northern New England features one of only two groupings of three contiguous states that have no city over 100,000 in population.[9] This is interesting and important because of the significance of scale. It is more difficult for large communities to focus on reorienting themselves to sustainable human and natural communities than it is for smaller communities. Hence, rural Vermont/northern New England is a likely place for bioregionalism to take root and hold.

To reinvigorate democracy in Vermont by focusing more authority and responsibility at the local level, Frank Bryan and John McClaughry write: "Representation is founded on citizenship [reared in] real polities: places where community and politics meet, where individuals learn the *habit* of democracy face to face, where decision making takes place in the context of communal interdependence" (1989: 3). Bryan and McClaughry propose developing a system of shires, government units bigger than Vermont's 246 cities and towns but smaller than its fourteen counties. Towns would still constitute the fundamental unit of governance, but to these forty shires would devolve much of the power now located at the state level, in areas such as education and welfare. Most of the shires would have a population of 5,000 to 15,000. "Bioregional identity" would be one of the many factors used in determining the boundaries of these shires, but based on the examples discussed by Bryan and McClaughry, characteristics of the natural landscape receive short shrift compared to population nodes and patterns of human activity, and, most especially, existing town borders that do not reflect landscape boundaries such as watersheds or ecoregions. Nonetheless, this book is an exciting foray into reconstituting Vermont on a smaller scale, an approach that could be fine-tuned to have the shire boundaries related to natural features as well as cultural ones.[10]

One final point before turning to sketch out a set of bioregional layers:

Vermont, like everywhere else, is greatly affected today by activities that take place outside its borders. Through the early to mid-1800s, the human inhabitants of Vermont were largely self-sufficient. The Abenaki hunter–gatherers and the colonial subsistence farmers needed little from outside their community or farm. Hence, their lives were somewhat insulated from the goings on beyond Lake Champlain or the Connecticut River. Those times are long gone, as those living in Vermont are now subject to the whims of an international milk market and to exotic organisms such as zebra mussels and Dutch elm disease. This interdependence will only increase with agreements like NAFTA and technological developments such as the internet. Vermont exists within a global economy. Without serious attention to these trends toward globalization, moving in a bioregional direction in any given locale will have inconsequential results; it is a necessary but insufficient undertaking to change our way of dwelling in the world.

We need some borders for political association. Given the natural and cultural overlays identified for Vermont, which borders make sense? Bioregional overlays are not likely to replace Vermont or other political entities. Such a change, welcome though it might be, would not be the best place to put our energies. Rather, I agree with Snyder, who writes:

> I am not arguing that we should instantly redraw the boundaries of the social construction called California, although that could happen some far day. We are becoming aware of certain long-range realities, and this thinking leads toward the next step in the evolution of human citizenship on the North American continent.
>
> (1992: 67)

Bioregional borders should be understood as a way of helping human beings to conceive a new-old way of living with the land. Once we have grasped that new way of living, our political institutions – including boundaries – will follow.

I propose four levels of bioregional overlays, suggestive of a set of alternative political boundaries. These overlays will help us to refocus our energies based on the knowledge of existing natural and cultural systems and their borders; to build upon the bioregional initiatives already begun; and to nurture the recovering natural landscape in the region. The basic level is a small watershed or a segment of a larger watershed. I cannot offer specifics for all of Vermont, since the wisest scale and boundaries must be determined by people living in these watersheds. I can offer an example from my home in Bristol, though. The basic bioregion here might consist of the Lewis Creek, Little Otter Creek and New Haven River watersheds, draining an area of over 170,000 acres and with a population of about 20,000. These watersheds would be compact enough to make sense for Vermont, where human-scale society still exists. In other words, bioregional borders based primarily at the larger scales discussed below would reduce the quality of human community that already exists in Vermont. These smaller base units would serve as the building blocks for sustainable human and natural communities.

The next level would be the major drainage basins flowing into Lake Champlain, Lake Memphremagog, the Connecticut River and the Hudson River.[11] The smaller watersheds discussed above would nest into this layer of bioregion. These second-level watershed bioregions, though still relatively small, are too large by Vermont cultural standards to serve as fundamental units. For me to have my primary attachment to the Otter–Lewis Basin, for instance, would increase the scale at which I relate to people and landscape, not reduce it as bioregionalism should. In other words, determining the borders of a bioregion as a primary unit must take into account social and cultural factors as well as natural ones. Such natural borders must be respected, but so must the ways in which people relate to each other – culturally, economically and politically. Eventually, some of these second-tier basins might relate more to bioregions at the third tier, located beyond Vermont's borders; for instance, the Batten Kill–Waloomsac–Hoosic Basin connecting to the Hudson Valley bioregions of New York, the Vermont portions of the Deerfield and Green Basins connecting to Lower Connecticut Valley bioregions in Massachusetts, and the Lake Memphremagog Basin relating to St. Lawrence bioregions to the north, where most of the basin is located.

The third bioregional tier is at the larger watershed level. Almost all of Vermont will be part of the Champlain Basin or the Upper Connecticut Valley Basin. At this level larger concerns, such as air pollution, water pollution and regional transportation issues will be dealt with. Coordination must occur on these issues within these basins due to biophysical realities, regardless of past cultural borders. It might be that eventually these third-tier bioregions come to replace states, and Vermont is carved down the spine of the Green Mountains into Champlain and Upper Valley bioregions, joining with parts of New York and New Hampshire. It is true that these mountains were a formidable barrier in the past, and until the railroads and later advances in highways and telecommunications, Vermont was virtually divided down the middle. Yet, as the above discussion indicates, Lake Champlain is a significant geological boundary, was a significant boundary for Native American peoples, and is still a cultural boundary between New England and New York. Hence, it is difficult to envision a bioregional state based on the Champlain Basin for quite some time. The Connecticut River, though, does not have such cultural and natural meaning as a boundary. There were early efforts to create a state based on the basin, so this might evolve more easily in the future.

The final overlay of boundaries is based at the ecoregional province level. Though Bailey has now revised his map so that Vermont is essentially part of two ecoregions – the Laurentian Mixed Forest Province and the Adirondack–New England Province – in his original take all of Vermont was part of the former. This ecoregion, stretching from the Maritime provinces to northern Minnesota, features similar forest types, climate and many cultural activities. At this level, discussions regarding forest health, value-added woodworking, wild-land reserve design and alternative energy systems for cloudy and cold locations would be most fruitful.

To reiterate, these bioregional overlays would surely be placed over existing political overlays at first. They would parallel the current federalist political system in the US and Canada. If we evolved to replace current political boundaries with these bioregional alternatives, a few crucial differences are obvious. The basic bioregions would be the focal point for social and cultural activities, much more like ancient city-states than modern counties. The policy responsibilities at the ecoregion–national level would be much less than at the national level, and the US would be carved into a series of ecoregional arrangements. Such dramatic changes would necessitate critical understanding of foreign policy, civil rights and the place for minorities, all of which are beyond the scope of this chapter. What is certain is that since bioregional boundaries are social constructions, these boundaries would evolve and change with new information and values.

Concluding thoughts

There were three major points made in this chapter. First, bioregional thinking requires that we focus much of our attention on the natural landscape and its health. Vermont's natural community is a rewilding and recovering one that offers the possibility of reworking the human place in nature. Second, there are a wide variety of bioregional initiatives underway in Vermont and surrounding areas, ranging in scope from 26-million acres to fewer than 100,000 acres. These public and private watershed-based initiatives provide opportunities for people to think and to act bioregionally. Third, there is no one set of boundaries that are the correct ones for a particular people and place. Rather, we should not become overly committed to a new set of boundaries to replace an old set of boundaries.

What does this discussion of bioregional opportunities in a small part of North America offer others? Vermont, like the hinterlands in much of the eastern US, is undergoing an unplanned but providential rewilding. Such events can happen anywhere in the world, and we should be ready for them. They present an important opportunity for a culture to reinhabit a landscape, and an opportunity to integrate a bioregional approach with physical and spiritual restoration in the places where we live. One of the reasons this rewilding has occurred is that Vermont is embedded in a postindustrial context. While this has been wonderful from the perspective of the Vermont landscape and for many of those living here, we must not forget that recovery and restoration are often related to increased exploitation elsewhere. I would like to believe that those living in Vermont are not as materialistic as Americans generally, but our consumption has not declined in any meaningful way. So, we get our food, wood and energy from "elsewhere" as our forest returns. But what is happening in that elsewhere? What is our responsibility for problems caused beyond our bioregion by our consumption and production? Vermont faces innumerable threats from beyond its borders. Since Vermont is part of the global economy, it has been tremendously influenced by forces from beyond, be it the demand for beaver pelts in the 1600s, the economics of wool trade and tariffs in the 1800s, or the politics and economics of dairy production. Human population is another major force affecting Vermont. With more people on the planet, demand for Vermont resources will increase and, more importantly, more people will move there, perhaps fundamentally altering the emerging cultural and natural synergy. Yet another set of external threats affect the natural systems, be it acid rain generated in the midwestern US or exotic species wreaking havoc in the forests and lakes. As the bioregional movement moves forward, it must fully engage these problems of globalism.

Despite these caveats, it is this new-old way of living in the world, I think, that is the key to bioregional possibilities in Vermont and elsewhere. Although having government engage in bioregional initiatives such as the Northern Forest Lands Study and the Lake Champlain Basin Program is important, reorienting the thinking of individuals is the key to a bioregional future. Government initiatives for regionalization and watershed-based ecosystem management must be supported by a cultural sensibility and respect for the landscape and place. Changing political institutions and economic systems will be very difficult, especially since these institutions and systems are moving in a direction of increased globalism. Leadership must come from below, since state and national governments are often threatened by the mere thought of bioregionalism.

In addition to altering people's way of thinking through bioregional initiatives that allow them to revisualize their place, educating our children is also crucial. For instance, one of my colleagues, John Elder, has undertaken a bioregional education project that teams Middlebury College students with teachers to develop bioregional-based curriculums at local schools. Spreading this approach across Vermont and the world is necessary if we are ever to get people to view the world from a bioregional perspective. If you can change enough people's way of thinking and living in place, then bioregional institutional change might follow.

Acknowledgments

This work draws on *Defining Vermont: A Natural and Cultural History* (University Press of New England, forthcoming), written by Stephen Trombulak and myself. I thank David Brynn, John Elder, Steve Trombulak and Mike McGinnis for their help in revising this chapter.

Notes

1 The leadership role in planning for more wild areas in Vermont and northern New England – based on conservation biology and existing land-use patterns – is held by the Greater Laurentian Region Wildlands Project and the Northern Appalachian Restoration Project (Sayen 1995).

2   The NFLC did suggest that the states should continue to coordinate their activities, and the Northern Forest Stewardship Act, passed by the Senate but not the House in 1996 – called for federal–state coordination on Northern Forest concerns. Nonetheless, this regional coordination cannot replace the value – including great symbolic value – of a regional institution like the NFLC. Furthermore, Maine and New York have not created such roundtables. New Hampshire has created a new roundtable, while Vermont has resurrected the preexisting Forest Resources Advisory Council.

3   The final draft of the Lake Champlain Management Plan went through its public review in August and September 1996. The seven major sub-basins are: Poultney–Metawee/South Basin, Otter/Lewis Basin, Winooski Basin, Boquet/Ausable Basin, Lamoille/Grand Isle Basin, Saranac/Chazy Basin and Missisquoi Basin.

4   Efforts have also been made to improve the quality of life for those living within the park. For instance, the Commission on the Adirondacks in the Twenty-First Century's report included recommendations to improve jobs, housing, health and education within the park (1990).

5   In 1977, Vermont passed enabling legislation for the formation of municipal conservation commissions. Approximately sixty-five towns – one-quarter of Vermont towns – have established such commissions. Though they are not bioregional, their local focus has been useful in focusing citizen attention on the role of local communities in environmental and conservation issues. This is not a trivial thing in Vermont, with its long tradition of town-meeting government. Such commissions can only help in shifting thinking in a bioregional direction.

6   For a detailed discussion of the natural communities of Vermont, see Thompson 1996.

7   Cultural boundaries are much more amorphous than natural ones. New England, for instance, has one of the clearest identities of any region in the country, and its border with New York – which has never been considered part of New England – is clear. There are significant historical differences on either side of the boundary. New York's colonial land scheme, for instance, was much more aristocratic than that used in New England. New York has been viewed as the center of cosmopolitanism in the nation, New England as the center of puritanism. In Vermont, snug against this cultural border, there are clear signs of ambiguity about its relationship to New York. Western Vermont milk goes to New York City, not to Boston; the football team of choice in the state is the New York Giants, not the New England Patriots; news coverage on the television stations in the state (based in Burlington) features stories about the Adirondacks across the lake more than they do about New England; and more and more people read the *New York Times* rather than the *Boston Globe*.

8   For more on the history of the interaction of the native population, the colonists and the landscape in New England, see Cronon (1983) and Merchant (1989).

9   The other is North Dakota, Montana and Wyoming. These figures are based on the 1990 Census (*Information Please Almanac* 1996: 748–82).

10   One weakness of this book – and it is a significant one – is that the authors are blind to the powers of unchecked capital. That is, they rightly identify the threat of big government to democracy and identify options to reduce big government and reinvigorate local democracy, but they do not discuss how small towns – or even states – will be able to deal with large corporations that are even less constrained than they are now due to the diminution in power of national governments.

11   These basins are:

   1   draining into Lake Champlain: Missisquoi Basin, Lamoille Basin, Winooski Basin, Otter–Lewis Basin and Poultney–Metawee–South Basin;
   2   the Lake Memphremagog Basin;

3   draining into the Connecticut River: Nulhegan Basin, Passumpsic Basin, Wells–Ompompanoosuc Basin, White Basin, Ottauquechee–Black–Saxtons Basin, West Basin and parts of two basins primarily in Massachusetts (Deerfield Basin and Green Basin); and

4   flowing into the Hudson River: the Batten Kill–Waloomsac–Hoosic Basin.

Interestingly, Vermont has just introduced new conservation license plates, with the post-administrative funds split between the Nongame Wildlife Fund and the Watershed Management Account. This account will make grants to community-based watershed projects in these four major watershed basins.

References

Bailey, R.G. (1995) *Description of the Ecoregions of the United States*, 2nd edn., Washington, D.C.: USDA Forest Service.

Bailey, R.G. and Cushwa, C.T. (1981) "Ecoregions of North America" (map), Washington, D.C.: US Fish and Wildlife Service.

Bryan, F. and McClaughry, J. (1989) *The Vermont Papers: Recreating Democracy on a Human Scale*, Chelsea, VT: Chelsea Green.

Collins, J. (1994) "The Adirondack Park: How a Green Line Approach Works," in C.M. Klyza and S.C. Trombulak (eds.) *The Future of the Northern Forest*, Hanover, NH: University Press of New England.

Commission on the Adirondacks in the Twenty-First Century (1990) *The Adirondack Park in the Twenty-First Century*, Albany, NY.

Cronon, W. (1983) *Changes in the Land: Indians, Colonists and the Ecology of New England*, New York: Hill & Wang.

—— (1996) "The Trouble with Wilderness; or, Getting Back to the Wrong Nature," *Environmental History* 1: 7–28.

Harper, S.C., Falk, L.L. and Rankin, E.W. (1990) *The Northern Forest Lands Study of New England and New York*, Rutland, VT: USDA Forest Service.

Haviland, W.A. and Power, M.W. (1994) *The Original Vermonters: Native Inhabitants, Past and Present*, revised edn., Hanover, NH: University Press of New England.

Henzel, L. (1996) Coordinator, Lewis Creek Association, personal communication, September.

Information Please Almanac (1996), New York: Houghton Mifflin.

Klyza, C.M. (1994a) "Lessons from the Vermont Wilderness," *Wild Earth* 4: 75–9.

—— (1994b) "The Northern Forest: Problems, Politics and Alternatives," in C.M. Klyza and S.C. Trombulak (eds.) *The Future of the Northern Forest*, Hanover, NH: University Press of New England.

Lake Champlain Basin Program (1994) *Opportunities for Action: An Evolving Plan for the Future of the Lake Champlain Basin*, Grand Isle, VT: Lake Champlain Basin Program.

Mad River Valley Planning District and Friends of the Mad River (1995) *The Best River Ever: A Conservation Plan to Protect and Restore Vermont's Beautiful Mad River Watershed*, Waitsfield, VT: Mad River Valley Planning District and Friends of the Mad River.

McKibben, B. (1995) *Hope, Human and Wild*, Boston, MA: Little, Brown.

Meeks, H.A. (1986) *Vermont's Land and Resources*, Shelburne, VT: New England Press.

Merchant, C. (1989) *Ecological Revolutions: Nature, Gender and Science in New England*, Chapel Hill, NC: University of North Carolina Press.

Morgan, K.O., Morgan, S. and Quitno, N. (eds.) (1994) *State Rankings 1994: A Statistical Overview of the 50 United States*, Lawrence, KS: Morgan Quitno.

Northern Forest Lands Council (1994) *Finding Common Ground: Conserving the Northern Forest*, Concord, NH: Northern Forest Lands Council.

Omernik, J.M. (1987) "Ecoregions of the Conterminous United States," *Annals of the Association of American Geographers* 77: 118–225.

Sayen, J. (1995) "A Proposal to Establish a Headwaters Regional Wilderness Reserve System," *Northern Forest Forum*, Headwaters Restoration Issue: 4–6, 8–10.

Smith, M.L., Carpenter, C., Fay, S., Burbank, D. and Burt, N. (1995) "Eco-Regions of New England and New York" (draft map), Durham, NH: USDA Forest Service.

Snyder, G. (1992) "Coming in to the Watershed," *Wild Earth*, special issue: 65–70; repr. in *A Place in Space: Ethics, Aesthetics and Watersheds*, Washington, D.C.: Counterpoint, 1995.

Thompson, E. (1996) *Natural Communities of Vermont: Uplands and Wetlands*, Montpelier, VT: Agency of Natural Resources.

Trombulak, S.C. (1994) "A Natural History of the Northern Forest," in C.M. Klyza and S.C. Trombulak (eds.) *The Future of the Northern Forest*, Hanover, NH: University Press of New England.

—— (1995) "Ecological Health and the Northern Forest," *Vermont Law Review* 19 (2): 283–333.

Trombulak, S.C. and Klyza, C.M. (1998, forthcoming) *Defining Vermont: A Natural and Cultural History*, Hanover, NH: University Press of New England.

Part II

Place, region and globalism

6 Bioregionalism, civil society and global environmental governance

Ronnie D. Lipschutz

Can bioregionalism play a role in the protection and conservation of the global environment? If by "bioregion" we mean a physically bounded ecosystem that constrains human social action and economy, the answer is probably "no." If, however, we use the term to denote a place or community, linked to nature, and with which residents identify in historical, cultural and material terms, our answer might well be different.

This chapter addresses the relationship between the bioregion as a functional and cultural entity and the possibilities of global environmental governance. I begin with a brief discussion of the relationship between the local and the global, and the functional difficulties that the state faces in trying, from a distance, to legislate and implement action. I then turn to an examination of the relationship between local political economy, the identity and forms of bioregionalism, and the possibility of global environmental governance. I conclude with an analysis of the political role of actors in specific locations who, as members of an emerging "global civil society," are taking on local environmental governance roles, and I consider their relationship to political economies and bioregionalism.

Generally, environmental degradation is a product of localized and bounded political economies and histories that are often dependent on biology and geography. If we wish to address the fundamental causes of environmental damage, it is with these political economies and histories, and their cultural and social attributes, that we must work. It is from these localized political economies that we must build a framework for addressing problems in what we call the "global environment." What we choose to call these regions is less important than that we choose to work within them.

In this chapter, I define "knowledge" as a system of conceptual relationships – both scientific and social – that explains cause and effect, and offers the possibility of human intervention and manipulation in order to influence or direct the outcomes of certain processes. "Local knowledge" encompasses such knowledge, as well as the specific and *sui generis* social and cultural elements of bounded social units. "Civil society" includes those political, cultural and social organizations of modern societies that are autonomous of the state, but part of the mutually constitutive relationship between state and society. "Global civil society" extends this concept into the transnational realm, where it constitutes something along the lines of a "regime" composed of local, national and global non-governmental organizations. "Governance" is, in Ernst-Otto Czempiel's words, the "capacity to get things done without the legal competence to command that they be done" (1992: 250). In this sense, it is a form of "authority" rather than jurisdiction (Lipschutz with Mayer 1996: ch.3). Finally, I use the term "political economy" in the nineteenth-century sense, as involving relations of power, wealth and production as well as history, culture and politics.

Local resources, local regimes

The ways in which we conceptualize environmental problems have a great deal of influence on how we try to address them. This may seem obvious, but the choice of one analytical framework over another can mean the difference between success and failure when the time comes to implement policy. Global environmental degradation is especially problematic in this respect. It is commonly viewed in terms of its most publicized manifestation – global warming (see Chapter 8 by Feldman and Wilt). But global warming is a secondary consequence of a variety of localized human activities and impact that, in turn, result in other forms of localized and regional environmental damage as well. The web of causes and effects we see is the result not of relatively straightforward physical relationships (e.g. rising temperatures leading to rising sea levels) but rather the interaction of complex social and natural ones (e.g. land-use, settlement patterns and class relations that lead people to live on low-lying islands in the Bay of Bengal, rendering them vulnerable to floods and storms). The public-policy literature generally fails to pursue these latter notions of complexity, even though they may have serious implications for policy-making and should be addressed prior to the negotiation of international agreements or conventions; the geography and history literatures that do address such complexity are often ignored by policy-makers.

This failure is understandable: The complexity of such sociospheric and biospheric connections produces a confusion of causes, consequences and linkages that are difficult to parse. More importantly, the complex linking of local and global means that some of the most important causes and consequences of global change – the sum of personal and collective choices and actions, and the cumulative effects on human lives – are inevitably distributed in an uneven fashion over a large number of bounded nation-states, cultures and societies, all of which complicates problem-solving. But the very real existence of complex social linkages underlines a fundamental problem in thinking about approaches to environmental protection: If existing "borders" are a problem, where and how are we to draw boundaries for managing "resources" so as to facilitate such protection and prevent damage?

We take it for granted that ecosystemic boundaries have little correspondence to political, economic and social institutions at the international level; indeed, this is one reason why international cooperation is seen as being so critical if global environmental problems are to be addressed. But the same poor fit is true at the national and even local levels: For historical and economic reasons, the jurisdiction of virtually all governments matches poorly to nature. This suggests that environmental governance is problematic no matter where one looks. What are we to do?

The institutionalization of "bioregionalism" on a worldwide scale is one potential solution: Supporters generally suppose that this would force governance to conform to nature. But on closer analysis we find that even bioregions are social constructs, inasmuch as regions in the natural world are distinguished by borderlands or ecotones rather than "hard and fast borders" (in fact, such fixed borders are largely an invention of the twentieth century; most cultural borderlands, such as that in the American Southwest, have this character). Imagine the kinds of political designs that would be required to adapt existing institutional frameworks, or create new ones, in the effort to replace counties, states and provinces by bioregions. We would, in such a process, be producing new situations in which history, economy and nature would (again) be mismatched.

This chapter focuses on something different: The relationship between specific human systems of resource exploitation – which included history, economy, culture and social relations – and the resources themselves; in other words, the development of a political economy of nature. Although the *physical* environmental effects of certain activities may be manifested or mediated via open access resource commons,[1] the activities contributing to these impacts tend to be bounded in social, economic and even physical terms. Global climate change is a phenomenon of the open access atmosphere; the production of greenhouse gases is a result of specific practices in specific places. As Ronald Herring points out: "[A]ll local arrangements for dealing with natural systems are embedded in a larger common interest defined by the reach of eco-systems beyond localities" (1990: 65). We can thus envision local systems of production and action – the immediate sources of environmental damage – as being nested within larger ones. These local "resource regimes,"[2] in turn, are part of economic, cultural and social networks of resource users and polluters rather than being either discrete or totally aggregated arrangements. Such user networks, moreover, are embedded in overlapping – but not necessarily coterminous – social, political, economic and physical spaces.

Resource regimes constitute only a part of the material base of a community. Of more consequence is that they are not only material, but also ideational, involving collective cognition, ideas and explanations. These regimes place resources and nature in a particular relationship to a community, thereby helping to constitute the meaning of the resource as well as the identity of the community, both historically and in the present. To put this another way, the identities of logging and fishing communities are bound up with their relationships to the resource. Resource regimes are, consequently, determined not only by the material conditions of production, they are also a consequence of the means of social reproduction, as well as being integral to such reproduction. Such regimes can be changed by the depletion of a resource, but can also be altered via ideational and cultural redefinition.

From where do these regimes come? There is no reason to think that, *contra* the conventional wisdom about international regimes, social institutions are always – or even very often – the product of negotiated bargains among participants (Hechter 1990). Much recent scholarship has been focused on the deliberate construction of international regimes by states under conditions of anarchy.[3] Many resource-using social institutions are, however, the outcome of decades, or even centuries, of material production, ecological change, and social interaction (McEvoy 1986; Berkes 1989; Ostrom 1989).[4] They may, for example, be historical and cultural artifacts, arising out of long-held customs, the structure of society and the nature of the resource being managed. To be sure, such regimes are a reflection of power relations as they have developed within a society, but these are not wholly unfettered or one-way relations of power. Rather, it is their historical constitution within a society that legitimizes such power relations and regimes and often leads to the reification of the institutions themselves.

Therefore, if we regard international agreements to protect the global environment as representing the "peak regime," so to speak, in a system of many thousands of smaller-scale resource regimes, the difficulty of generating responsive action through these regimes from the top-down becomes immediately apparent. It is not even clear that international diplomatic efforts are linked to anything meaningful, since they cannot begin to alter the fundamental rules, roles and relationships that constitute these micro-level regimes. This may be the case not only in developing countries but in the industrialized world as well (Lipschutz with Mayer 1996: ch.2).[5] Fiscal and regulative levers may be able to reach into these institutions in some places, thereby modifying activities that are environmentally damaging, but they are likely to be next to useless in others. This raises the question: What then?

One answer is to reconceptualize our view of the relationship between society and nature and our understanding of the relative roles of agency and structure in these relationships, as I suggested above. This involves, as well, a better understanding of how our material surroundings are implicated in the constitution of community and identity, and vice versa. We are all, in many ways, part of and implicated in such relationships, even though we may be quite unaware of them. To change these relationships – to do so consciously and collectively – is to act in a way that not only changes our *material* relationship to nature but also *reconceptualizes* our place in nature. This involves "reconstructing" resource regimes in material terms, and in the ways our individual and collective identities are implicated in these regimes.

An illustration of this process can be seen in the Mattole River Watershed on the "Lost Coast" of California. The river rises in the forests at the base of the King Range, which runs parallel to the coastline, traverses a distance of not much more than sixty miles through the counties of Mendocino and

Humboldt, and reaches the Pacific Ocean not far from the small town of Petrolia, south of Cape Mendocino. The entire Mattole watershed of perhaps 300 square miles is inhabited by no more than 3,000 people. Historically, the economy of the valley has been based on resource extraction, logging, ranching, fishing and not much else, because it is relatively inaccessible due to the poor roads linking it to the rest of the coast. Apart from its beauty and tranquillity, the Mattole River Valley has acquired a reputation for the cooperative ventures of its citizens, ranchers, fishermen and environmentalists, who have been working together toward restoration of the river's largely depleted salmon runs.

In the early 1980s, a number of individuals, including several with technical backgrounds in ecology and fisheries, established the Mattole Watershed Salmon Support Group, a private, nonprofit organization with the goal of reviving the salmon runs through river restoration. The group went about its work very systematically, surveying populations, spawning grounds and nests as potential causes of fishery decline, and doing whatever it could to maintain or improve the salmon habitat in the river. This included operation of "homemade" hatching and rearing facilities. In the course of work on the fishery, it became clear that one of the major sources of damage was erosion and the deposition of sediment into the river via landslides, road maintenance and activities associated with ranching. Eventually, members of the Salmon Support Group found it necessary to work with ranchers, loggers and others in order to achieve their objectives. They joined together in the Mattole Restoration Council in order to control some of the more damaging activities. State and county authorities were also willing to support the project because it dovetailed with their goals. (For a discussion of the Mattole Restoration Council, see Chapter 12 by McGinnis, House and Jordan.)

While the salmon restoration effort has not yet proved definitively successful, some guess the project may take decades to succeed – the experience has been described as "transformative" for the residents of the valley. During the 1980s, there was endemic conflict among the different economic and social groups living near the Mattole. Working together toward a common goal of fishery restoration, and utilizing both local knowledge and scientific analysis, helped to minimize some of the fractures and frictions within the watershed community.

The Mattole project was initially conceived of in bioregional terms; several of its progenitors are well known in the North American bioregional movement. The focus of the ecological restoration effort, however, was not on the bioregion as a whole. Rather, the focus was on a resource – the salmon – within the system of production, consumption and community. The focus of regime reconstruction is not the Mattole watershed, but the political economy of human society within that watershed. In this instance, for geographic and other reasons, political economy and bioregion map closely on one another. In other contexts, this might not be so.

From global to the local

Such cases provide functional evidence that indicate that regime reconstruction at a local level is possible. Also, there are analytical grounds for a more localized approach, based on the structure and nested nature of resource regimes. These arguments rest on the notion that people are more likely to act collectively when their personal experiences and surroundings are implicated in a process than when they are expected to respond to government directives from a distance or abstract predictions of future dislocations.

More than this, the resistance of actors within social institutions to external directives makes change from a distance a difficult proposition (Wade 1987: 105), for a number of reasons. If we regard a resource regime as, first, the embodiment of power relations within a society and, second, having as one of its primary goals social reproduction over time, the potential for significant, large-scale reform, in the absence of major crisis, would appear to be severely circumscribed. Such institutions can be changed, as indicated by the history of environmental regulation over the past three decades, but opposition can be quite powerful as well.

This last point is illustrated in the United States by ranchers' opposition to increased grazing fees, farmers' antagonism toward sharing "their" water with wildlife, and loggers' intense dislike of environmentalists and Spotted Owls (similar cases can be found around the world).[6] What, after all, is a rancher without rangeland, a farmer without water, a logger without trees? A group of resource users acting through a resource regime is not merely an economic construct or, for that matter, one driven by the presumed logic of "rational choice"; rather, it is an historically constituted entity, a social institution.

Consequently, a resource regime is characterized by specific relations of power, wealth, legitimation and affection (or disaffection) that have developed over time. Within such a regime, individuals are linked together through bonds of social obligation, and their access to the resource is based upon patterns of access and historical distribution of resources (Ostrom 1990).[7] While these patterns may not be distributionally just, as something akin to social and natural contracts that bind people to their homes and communities[8] they do have the weight of history behind them.[9]

Changing the internal structural relationships, rules and practices within a resource regime means alterations in underlying property rights, a process that is only possible through a renegotiation of these rights (Nuijten 1992). Inasmuch as the process of establishing or changing a localized resource regime is fundamentally political, it is also messy (Stone 1988: conclusion). Legislation originating from "above" is rarely able to take into account the valid concerns of all stakeholders in a resource because of a lack of information about the institutional history and path dependency[10] of a resource management system, which are of critical importance to its revision (McEvoy 1988). Moreover, the obstructions to altering this pattern can be significant; sunk costs are high and institutionalized paths are difficult to renegotiate (Libecap 1989). Much of the appropriate knowledge required to change these patterns is only available locally. Resource regimes, in being created, reconstituted, or revised, *must* take into account contextual knowledge and local conditions if they are to have any chance of functioning in a sustainable fashion.

A similar caution applies at the international level: Collective action emerges only when someone's interests appear to be threatened by changes in the *status quo*. Environmental regimes are negotiated by states, which are highly resistant to imposing on themselves an enforceable obligation to alter domestic social institutions in a serious way (the "two-level game" problem; see Evans, Jacobson and Putnam 1993). Indeed, this is why so much attention is paid to economic "incentives" that alter relative prices as a means of changing consumer behavior on a large scale. Not only are such incentives naturalized by reference to market "efficiency," they leave untouched fundamental structures and relationships within society. In reproducing social institutions through legislative and fiscal mechanisms, one is, in essence, recreating the relationships that caused the environmental degradation in the first place. In many instances, they impose costs on those who are badly placed to challenge the reforms because of lack of political power and wealth. In other words, we can see regimes and social institutions as mechanisms intended to maintain and/or restore structural relations of power as arrangements to facilitate collective action and cooperation (as is commonly assumed).[11]

Consequently, the analytical argument for focusing locally rests on five points: (1) scale of ecosystems *and* resource regimes; (2) assignment of property rights; (3) availability and location of social knowledge; (4) inclusion of stakeholders; (5) sensitivity to feedback. These are discussed in greater detail elsewhere (see Lipschutz with Mayer 1996: ch.2), but the essential point is that these elements are all part of the political economy and history of a localized resource regime, and necessary to its long-term sustainability.[12]

How might such changes be operationalized and implemented? Interestingly, I would argue, they are happening around the world, but not as a result of international or national initiatives. Rather, the reconstruction of local resource regimes is one type of agent-based response to global structural change and the weakening of state authority and capabilities. It is to this topic that I turn next.

Governance and the environment

As we approach the end of the twentieth century, there is, inevitably, a great deal of speculation on the future of human civilization and politics, ranging from the apocalyptic (Kaplan 1996) to the optimistic (Ausubel *et al.* 1996) to the banal (Huntington 1993). Some have suggested that we confront a "new mediaevalism"; others have proposed as organizing principles "heteronomy" or "heterarchy," a system in which political rules are dispersed among different types of functional jurisdictions, operating at local, national and global levels, or across one or more of them.[13] In discussing the first of these three concepts, Ole Wæver argues that

For some four centuries, political space was organized through the principle of territorially defined units with exclusive rights inside, and a special
kind of relations on the outside: International relations, foreign policy,
without any superior authority. There is no longer one level that is clearly
the most important to refer to but, rather, a set of overlapping authorities.

(1995: n.59)

What is important here is the concept of *authority* – in the sense of the ability to
get things done because of one's legitimacy, as opposed to one's ability to apply
force or coercion – rather than "law" or "power." As John Ruggie points out
(1989: 28), in a political system – even a relatively unsocialized one – *who* has
"the right to act as a power" (or an authority) is at least as important as the
capability of actors to force others to do their bidding.

In this emerging "heteronomy," political authority will be dispersed
among many centers of jurisdiction, often on the basis of specific issues
rather than territories. In a way, this reorganization of authority will generate
a form of multilevel functionalism (really, functional differentiation) rather
than federalism, inasmuch as different authorities – NGOs, local governments, corporations, even churches and schools – will deal with specific
problems, some spatial, others not, such as toxic wastes moving through a
neighborhood here, protection of a marsh there, monitoring of water quality
elsewhere. These types of matters will remain embedded within a global
economic system, and will come under general legitimizing regulations. It
will be under the direct governance of the specific responsible authority. Such
functionalism will reach beyond localities into and through the global system
via networks of knowledge, practice and norms. Nevertheless, it will be
rooted in locale.[14]

This is not the same as the functionalism of the 1960s. Whereas the theories
of Mitrany (1966) and Haas (1964) envisioned political integration as the
outcome of international functional coordination, it is likely that contemporary
functionalism may lead to something quite different, a consequence of the
marriage of local knowledge and governance at multiple levels. In the present
instance, functionalism can be understood as a consequence of rapid social
innovation, of the generation of new scientific-technical and social
knowledge(s) required to address different types of contemporary issues and
problems.[15] Inasmuch as there is too much scientific and social knowledge for
any one actor, whether individual or collective, to assimilate, it becomes necessary to establish knowledge-based alliances and coalitions whose logic is only
partly based on space or, for that matter, hierarchy. "Local" knowledge is
spatially-situated while "organizational" knowledge – how to put knowledge
together and use it – is not bound to place, although successful organization
aimed at solving a localized functional problem must nonetheless be based on a
solid understanding of local social relations (Mayer 1995). Together, the two
become instrumental to technical and social innovation.

In the environmental arena, such arrangements are represented by the

Global Rivers Environmental Education Network (GREEN), which has projects in 136 countries, and the River Watch Network (RWN), based in Vermont. GREEN:

> seeks to improve the quality of watershed and rivers, and thereby the lives of people. GREEN uses watersheds as a unifying theme to link people within and between watersheds. . . . Each watershed project is unique, and how it develops depends upon the goals and situation of the local community. . . . As they share cultural perspectives, students, teachers, citizens and professionals from diverse parts of the world are linked by a common bond of interest in and concern for water quality issues.
>
> (cited in Lipschutz with Mayer 1996: 158)

RWN is more focused on the linkages between technological and scientific competence and political action, without much reference to larger goals. As RWN's materials put it, "We can help you clean up your river":

> River pollution . . . is generated by all of us and its solution requires active citizen participation. Federal, state and local governments are frequently unable to tackle these water quality problems because their resources for river monitoring are severely limited. . . . Gathering and interpreting scientifically credible water quality data underlies every River Watch effort. . . . RWN will never just send you a kit with a page of instructions for water sampling. . . . Each River Watch program is individually designed to meet the particular needs of its community and the conditions of its river. . . . RWN staff are river experts *and* community organizers.
>
> (cited in Lipschutz with Mayer 1996: 158; emphasis in original)

Both networks are only part of a growing worldwide effort to protect and restore river and stream watersheds (Coate, Alger and Lipschutz 1996).

Acquisition of such knowledge and practices leads to new forms and venues of authority, in that only those with access to such capabilities can act successfully. In some sense, the "management" function locates itself at that level of social organization at which the appropriate combination of local and global knowledges come together (Lipschutz with Mayer 1996: ch.2). This level is more likely to be local – in the lab, the research group, the neighborhood, the watershed – than global. Or, as Richard Gordon puts it (albeit on the subject of technological innovation):

> Regions and networks . . . constitute interdependent poles within the new spatial mosaic of global innovation. Globalization in this context involves not the leavening impact of universal processes but, on the contrary, the calculated synthesis of cultural diversity in the form of differentiated regional innovation logics and capabilities. . . . The

effectiveness of local resources and the ability to achieve genuine forms of cooperation with global networks must be developed from within the region itself.

(1995: 196, 199)

But such functionalist regionalization points back toward the problem of political fragmentation alluded to above: Lines must be drawn somewhere, whether by reference to nature, power, authority or knowledge. From a constructivist perspective, such lines may be as "fictional" as those which currently separate one country or county from another. Still, they are unlikely to be wholly disconnected from the material world, inasmuch as they will have to map onto already-existing patterns and structures of social and economic activity. Inasmuch as the sources of environmental degradation are, more often than not, rooted in the political economies of places, rather than in a bioregion, working to change the relations and practices inherent in those local political economies is central to protecting nature. Hence, it makes sense to first draw lines on the basis of those political economies; bioregional lines can come later.

Global civil society and environmental governance

The logic of local rule and authority discussed above is not merely theoretical; as I suggested above, it is being implemented globally, in many places, by what I call "global civil society," a transnational formation of primarily non-governmental organizations that is functionally place-based but normatively global (Lipschutz [1992] 1996; Wapner 1996). Global civil society encompasses actors engaged around many different issue areas, sometimes working with national governments and international institutions, sometimes opposing them. This is especially true in the environmental movement.

Some of the actors in these networks and coalitions are engaged in fairly-localized projects aimed at protection, conservation or restoration of ecosystems, habitats and watersheds, some have a more urban focus and are concerned with environmental justice, pollution, health, transportation and so on, and others are more global in scope, through transnational networks or educational projects directed toward the restoration of rivers and streams. These kinds of activities have been extensively documented elsewhere (Lipschutz with Mayer 1996; Wapner 1996). Below, I consider the relationship between this "global civil society" and "environmental governance."

Much can be learned about this relationship by examining the histories and political economies of environmental projects at the local and regional level. One example can be found in John Walton's study (1992) of the century-long struggle against Los Angeles by the residents of the Owens Valley in eastern California. He describes how local groups, engaged in resistance against the state, did not manage to achieve success in their efforts until they were able to draw on the expanding authority of the federal state, and the legitimation of various environmental strategies, as a means of putting pressure on LA to alter its patterns of water removal from the Valley.

In a broader sense, this coalition took advantage of local knowledge, and a nationally redefined ideology of ecology legitimated by the US federal government, to recast social meanings for political purposes. Whereas LA tapped the Owens Valley water sources for industrial and urban growth, both the local residents and resource managers in Washington, D.C., sought to conserve water and restore the landscape by framing the conflict in terms of an increasingly accepted story of environmental protection and restoration. This is not to suggest that self-interest was absent from the story; only to point out that collective action required collective meanings. There is another important lesson to be gleaned from this example: while it is difficult to expect people to act collectively to protect things that are abstract – such as the changing atmosphere – their behavior toward things that are concrete and local can have important consequences for protecting abstract things. The experience of collective action on behalf of the local environment can serve to instill an ethic that will apply outside of that locality.

Beyond this, the insight provided by Walton's story is that our conventional concepts of the state and governance are too limited. The state – even a federal one – is not restricted to discrete levels of government; it is more than that. As Theda Skocpol points out:

> On the one hand, states may be viewed as organizations through which official collectivities may pursue collective goals, realizing them more or less effectively given available state resources in relation to social settings. On the other hand, states may be viewed more macroscopically as configurations of organizations and action that influence the meanings and methods of politics for all groups and classes in society.
>
> (1985: 20)

Skocpol offers a conception of the state that is too broad in encompassing society, but her point is, in my view, an important one. The state is more than just its constitution, agencies, rules and roles, and it is embedded, as well, in a system of governance. James Rosenau argues that:

> Governance . . . is a more encompassing phenomenon than government. It embraces governmental institutions, but it also subsumes informal, non-governmental mechanisms whereby those persons and organizations within its purview move ahead, satisfy their needs, and fulfill their wants. . . . Governance is thus a system of rule that is as dependent on intersubjective meanings as on formally sanctioned constitutions. . . .
>
> (1992: 4–5)

From this view, state and civil society are mutually constitutive and, where the state engages in *government*, civil society plays a role in *governance*. What is striking, especially in terms of relationships between environmental organiza-tions and institutionalized mechanisms of government, is the growth of institutions of governance at and across all levels of analysis.[16]

From a bioregional perspective, moreover, this growth of governance is central, for there is, in principle, no reason why bioregional governance cannot coexist with, supplement or, eventually, supplant contemporary units of govern-ment. Change must first come within existing institutions, but as change takes hold, some of their functions can be taken over through bioregional gover-nance, as the organizations associated with the Monterey Bay Sanctuary have begun to do. Indeed, counties and cities would probably be only too glad to transfer such responsibilities (Lipschutz 1997).

Even though we should recognize that there is no world government as such, there is an emerging system of *global governance*. Subsumed within this system of governance are both institutionalized regulatory arrangements – some of which we call "regimes" – and less formalized norms, rules and procedures that pattern behavior without the presence of written constitutions or material power.[17] This system is not the state, as we commonly understand the term, but it is state-like in Skocpol's second sense. Indeed, we can see emerging patterns of behavior in global politics very much like those described by Walton in the case of the Owens Valley: alliances between coalitions in global civil society and the international governance arrangements associated with the UN system.[18] Each of the actors, at one time or another, finds it useful to ally with others, at other levels, so as to put pressure on yet other actors, at still other levels. The result might look more like a battlefield than a negotiation – and, indeed, violence is an all-too-real component of this particular campaign – but, although there is no definitive ruler, the process is not entirely without rules or structure.

To push this argument further, let us return, for a moment, to what scholars of international environmental policy and politics regard as the *sine qua non* of their research: The fact that environmental degradation respects no borders. This feature thrusts many environmental problems into the international realm where, we are reminded, there is no government and no way to regulate the activities of sovereign states. From this follows the need for international coop-eration to internalize transboundary effects, a need that leads logically to the creation of international environmental regimes. Such regimes are the creation of states, and scholars continue to argue about the conditions necessary for their establishment and maintenance. Whether they undermine the sovereignty of states or are, in themselves, a form of state-building is, as yet, unclear (Deudney 1993; Thompson 1992); what is less well-recognized and acknowl-edged is that some regimes are merely the "tip of the iceberg" of necessary action, or they will be if they reach fruition.

Much of the implementation and regulation inherent in regimes will have to take place at the regional and local levels, in the places where people live, not where their laws are made. If international regimes are to be successful – what-ever "success" means in such a context – they must, for all practical purposes, function as global institutions of governance with elements at the local, regional, provincial, national and international levels. They will, in effect, transfer some of the jurisdictional responsibilities of the state both upwards and downwards, enhancing political authority at the global as well as the local levels. People will find it necessary to come to terms with the changes required to deal with these global problems, through local law, norms, custom and action, just as water quality and river protection are mandated on high but implemented within the watershed.

Such governance is characteristic of the emerging global political economy characterized by economic integration and political fragmentation. The fundamental units of governance are, in this system, defined by both *function* and *social meanings*, anchored to particular places, but linked globally through networks of knowledge-based relations. These relations develop when the costs of acquiring information through "normal" channels becomes too great. These relations bear a remarkable resemblance to transactions and economies oriented around kinship relations – and bioregionalism too – in which trust and membership replace formal hierarchies and markets (Ouchi 1980; Alvesson and Lindkvist 1993). The phenomenon of "networking," characteristic of relations within global civil society, resembles this form of organization. It is a form that lends itself to cooperation without centralization, without "global management."

A governance system composed of collective actors at multiple levels, with overlapping authority, linked together through various kinds of networks – a heterarchy – might be as functionally efficient as a highly centralized one. Such a decentralized system of governance has a number of advantages over a real or imagined hierarchical counterpart. As Donald Chisholm points out:

> [F]ormal systems often create a gap between the formal authority to make decisions and the capacity to make them, owing to a failure to recognize the necessity for a great deal of technical information for effective coordination. Ad hoc coordinating committees staffed by personnel with the requisite professional skills appear far more effective than permanent central coordinating committees run by professional coordinators.
>
> (Chisholm 1989: 11)

Chisholm goes on to argue that formal systems work so long as appropriate information, necessary to the function and achievement of their goals, is available. The problem is that:

> Strict reliance on formal channels compounds the problem [of trying to prevent public awareness of bureaucratic failure]: reliable information will not be supplied, and the failure will not be uncovered until it is too late to compensate for it. Informal channels, by their typically clandestine nature and foundation on *reciprocity and mutual trust*, provide appropriate means for surmounting problems associated with formal channels of communication.
>
> (Chisholm 1989: 32; emphasis added)

Compare this observation with Richard Gordon's discussion of the organizational logic of innovation:

> While strategic alliances involve agreements between autonomous firms, and are oriented towards strengthening the competitive position of the network and its members, inter-firm relations *within* the alliance itself tend to push beyond traditional market relations. Permanently contingent relationships mediated by strict organizational independence and market transactions – the arms-length exchange structure of traditional short-term linkages – are replaced by long-term relations intended to endure and which are mediated by highly personalized and detailed interaction. . . . *Cooperative trust, shared norms and mutual advocacy overcome antagonistic independence and isolation.*
>
>   (Gordon 1995: 183–4; first emphasis in original; second emphasis added)

Conclusion

Whether the units of governance discussed here will be bioregions will be largely a matter of contingency and context. Where organized groups of people are engaged in local environmental governance, they often adopt the language of bioregionalism in order to distinguish their activities from those of counties, cities and states (Lipschutz with Mayer 1996: ch.4). But it makes little sense to be dogmatic about this: The logging company or rancher with property on both sides of a ridgeline is not going to observe different rules in the two adjacent watersheds.

It might be better to think in terms of the relationship between the material base of a place and its meanings to those who live there. Constructing meanings for places as the basis for the social reconstruction of resource regimes creates committed "communities of place" among those who participate in this exercise. Such meanings help to transform a locale from one fragmented into multiple private and public tracts to one held in collective trust by a newly-sensitized community. Within this new structure of meaning, the community acquires and expresses a stake in the place and, to some degree, becomes dependent on the protection and maintenance of place.

Community and place become mutually constitutive, the identity of the former based on the sustenance of the latter, rather than locked in a relationship of extractive exploitation and heedless degradation. In this process, a subtle transformation of property rights has taken place, even though formal ownership may not have changed. A form of common trust has been created, in which the community is granted a say in how that place is to be used (Ostrom 1990). Note that this transformation of meaning is neither a conflict-free process nor a process that ends conflict within a community. Changes in meanings do not automatically lead to immediate changes in the ownership or utilization of property. The process of renegotiation is – indeed, it must be – an ongoing one (Mitchell 1993: 112–3).

What does this all add up to? This "thing" that we call the "global environment" is, in many ways, a mosaic rather than a seamless picture. The mosaic adds up to a whole, but the whole is not dependent on every single piece of the mosaic being in place. The overall picture continues to remain apparent even when some of the pieces are missing (so the appropriate metaphor is, perhaps, hologram rather than mosaic). In saying this, I do not mean to suggest that some pieces of the whole are thoughtlessly expendable; rather, it is an acknowledgment that not all of the pieces of the whole can necessarily be saved or sustained. More than this, it is a recognition that there is no single place from which the pieces can be sustained; the responsibility lies, instead, with those who are able to find their "place" in each individual piece.

Given this metaphor – for that is what it is, of course – the role of global civil society and "bioregions" in environmental governance starts to become clearer. What is needed for global environmental sustainability is, on the one hand, a common project and, on the other, a recognition that the efforts in each piece of the mosaic must be sensitive to nature (and culture) in each piece of the mosaic. Governments are important – indeed, they are essential to the legitimation of local environmental governance – but most of them are not in a position to see these manifold projects through. Even county and municipal governments, as local as they are, are often not in a position to manage environmental restoration for fiscal reasons. The responsibility, ultimately, will have to rest on social institutions, such as common pool property resource systems or bioregions or restoration projects, developed and run by groups of stakeholders based in the "civil societies" of these many places (Lipschutz with Mayer 1996: ch.8).

In policy terms, what I have offered in this chapter is not a very satisfying or parsimonious framework. It does not provide an entry for either easy explanation or manipulation. It relies on the possibly heroic and hopeful assumption that people can and will help to create social-choice mechanisms (Dryzek 1987), in their collective self-interest, that may also help to protect nature. But global civil society does offer more than global management or the imposition of laws by a centralized government on a passive or resistant population. My framework suggests that people, acting locally, can have a real and significant global impact. Such social and political change will not occur quickly, nor will it come easily, and it will never encompass the entire world. But, at the very least, by illuminating and examining change where it is underway, we can offer to others a model of action based on local knowledge that can, over the longer term, make a meaningful difference.

Notes

1 Oran Young (1982) has used this term in an international context. I have adopted it here in recognition of the existence of resource-using social institutions at the local and national levels too.
2 The assumption of international "anarchy" is not necessarily a valid one, at least not as it is conventionally stated (see Onuf 1989; Bergesen 1990; and Wendt 1992). On the possibility of nonstate regimes, see Conca (1996).

3 This does not mean that they are not "negotiated" in the sense of being dynamic and flexible; only that there are no formal negotiations involved. See Tsing (1993) for a discussion of this concept of negotiation.

4 Indeed, if one adopts some of the approaches of recent sociological research, the "encounters" between groups such as environmentalists and loggers or ranchers bear a striking resemblance to " 'interface' situations [in the developing world] where the different life-worlds interact and penetrate" and in which "actors' interpretations and strategies . . . interlock through processes of negotiation and accommodation" (Long 1992: 5–6).

5 The point is that institutional reforms, such as the US Endangered Species Act of 1973, that ignore the histories and political economies of the communities affected by them, are almost certain to generate "resistance" (*The Economist* 1995: 21–2). Ironically, perhaps, such resistance is often viewed favorably when it occurs in developing countries, less favorably when it happens in industrialized ones.

6 Even patterns of resource-use systems in industrialized countries are the result not of rationalized economic planning but of historically contingent episodes and accretions. Libecap's account of mineral rights contracting in California and Nevada (1989: ch.3) nicely illustrates this last point.

7 I owe this formulation to a comment by Mike McGinnis.

8 Witness, for example, the many recent claims by Native Americans demanding the restoration of historical property rights to resources such as salmon, and the reaction of others to these demands. More to the point, what is the difference between "local" and "international" regimes if both are unjust and reproducers of inequitable power relations? This is a valid question for which I do not have an answer. In spite of the optimism and idealism of many proponents of local control, localism is not always a democratic force (Stone 1988: 300–4); indeed, it was the federal government's use of its leverage against "states' rights" that was responsible for the legislation of civil rights in the 1960s and 1970s.

9 "Path dependency" means that where you start strongly influences where you end up. The classic example is the QWERTY typewriter keyboard, which was originally designed to slow down typists. It is generally agreed to be an inefficient design, but the sunk costs of using it are, by now, much too high to write off. For a discussion of path dependency, see Krugman (1994).

10 The literature on the emergence of international regimes is divided on this point. Some realists see them as instruments of state power that emerge only at the behest of powerful or hegemonic states (Krasner 1983); others doubt whether they can have any effect at all on the conduct of international relations (Mearsheimer 1995). Liberal institutionalists are more optimistic on this point, and see possibilities for reform and maintenance (Keohane 1984). Clearly, regimes might not serve the interests of the powerful – as in the case of UNCLOS – but then their "effectiveness" is in doubt (Haas, Keohane and Levy 1993; Young 1994).

11 Although there are long-standing arguments for the decentralization of administrative systems and functions in development planning, the argument presented here has its roots in the "sociology of knowledge" literature. For a discussion of problems with decentralization, see Karim (1991). For a good summary of the sociology of knowledge literature, see Haas (1992: 20–6).

12 The best-known discussion of the "new mediaevalism" is to be found in Bull (1977: 254–5, 264–76, 285–6, 291–4). The notion of "heteronomy" is found, among other places, in Ruggie (1983: 274, n.30). The term "heterarchy" comes from Bartlett and Ghoshal (1990), cited in Gordon (1995:181).

13 One example of this is the growing environmental justice movement, which is becoming globalized and addressing not only the local disposition of toxic wastes but its export and disposal in other places around the world (for example, see Clapp 1994).

14 The following paragraphs are based on Gordon (1995). Gordon argues for the existence of three "logics" of world-economic organization: internationalization,
multi/transnationalization and globalization. The last is "heterarchical" and
nonmarket, and (as he puts it) involves "valorization of localized techno-economic
capabilities and socio-institutional frameworks . . . [with] mutual reciprocity between
regional innovation systems and global networks" (from "Concurrent Processes of
World-Economic Integration: A Preliminary Typology," handout in colloquium, 30
November 1994, at the University of California, Santa Cruz). Gordon is, essentially,
making arguments about the organization and flows of knowledges, that map rather
neatly (I think) onto global networks of civil society. I anticipated some of this in an
earlier unpublished paper (Lipschutz 1991).
15 For one perspective on this phenomenon, see Leatherman, Pagnucco and Smith
(1994: esp. 23–8).
16 This point is a heavily disputed one: To wit, is the international system so underso-
cialized as to make institutions only weakly-constraining on behavior (as Stephen
Krasner might argue), or are the fetters of institutionalized practices sufficiently
strong to modify behavior away from chaos and even anarchy (as Nicholas Onuf
might put it)? See Krasner (1993); Onuf (1989).
17 A good illustration of this process can be found in Wilmer (1993).
18 Two of the best-known works addressing regimes and the conditions of creation and
maintenance are Krasner (1983) and Keohane (1984). More recent works on environmental regimes include: Haas, Keohane and Levy (1993); Young and Osherenko
(1993); Young (1994); Litfin (1994).

Bibliography

Alvesson M. and L. Lindkvist (1993) "Transaction Costs, Clans and Corporate Culture,"
Journal of Management Studies 30 (3): 427–52.
Ausubel, J.H. *et al.* (1996) "The Liberation of the Environment," *Dædalus* 125 (3):
1–17.
Bartlett, C. and Ghoshal, S. (1990) "Managing Innovation in the Transnational Corpo-
ration," in C.Y. Doz and G. Hedlund (eds.) *Managing the Global Firm*, London:
Routledge.
Bergesen, A. (1990) "Turning World-System Theory on Its Head," in M. Featherstone
(ed.) *Global Culture: Nationalism, Globalization and Modernity*, London: Sage.
Berkes, F. (1989) *Common Property Resources: Ecology and Community-Based Sustainable
Development*, London: Belhaven Press.
Bromley, D.W. (ed.) (1992) *Making the Commons Work: Theory, Practice and Policy*, San
Francisco: ICS Press.
Bull, H. (1977) *The Anarchical Society: A Study of Order in World Politics*, New York:
Columbia University Press.
Chisholm, D. (1989) *Coordination Without Hierarchy: Informal Structures in Multiorga-
nizational Systems*, Berkeley: University of California Press.
Clapp, J. (1994) "The Toxic Waste Trade with Less-Industrialised Countries: Economic
Linkages and Political Alliances," *Third World Quarterly* 15 (3): 505–18.
Coate, R.A., Alger, C.F. and Lipschutz, R.D. (1996) "The United Nations and Civil
Society: Creative Partnerships for Sustainable Development," *Alternatives* 21 (1):
93–122.

Conca, K. (1996) "International Regimes, State Authority and Environmental Transformation: The Case of National Parks and Protected Areas," paper prepared for the 92nd annual meeting of the American Political Science Association, San Francisco.

Czempiel, E.-O. (1992) "Governance and Democratization," in J.N. Rosenau and E.-O. Czempiel (eds.) *Governance without Government: Order and Change in World Politics*, Cambridge: Cambridge University Press.

Deudney, D. (1993) "Global Environmental Rescue and the Emergence of World Domestic Politics," in R.D. Lipschutz and K. Conca (eds.) *The State and Social Power in Global Environmental Politics*, New York: Columbia University Press.

Dryzek, J.S. (1987) *Rational Ecology: Environment and Political Economy*, Oxford: Basil Blackwell.

The Economist (1995) "When Mining Meets Golf," July 1: 21–2.

Evans, P.B., Jacobson, H.K. and Putnam, R.D. (eds.) (1993) *Double-Edged Diplomacy: International Bargaining and Domestic Politics*, Berkeley: University of California Press.

Gordon, R. (1995) "Globalization, New Production Systems and the Spatial Division of Labor," in W. Litek and T. Charles (eds.) *The Division of Labor: Emerging Forms of World Organisation in International Perspective*, Berlin: Walter de Gruyter.

Haas, E.B. (1964) *Beyond the Nation-State*, Stanford: Stanford University Press.

Haas, P.M. (1992) "Introduction: Epistemic Communities and International Policy Coordination," *International Organization* 46 (1): 1–36.

Haas, P.M., Keohane, R.O. and Levy, M.A. (eds.) (1993) *Institutions for the Earth: Sources of Effective International Environmental Protection*, Cambridge: MIT Press.

Hechter, M. (1990) "The Emergence of Cooperative Social Institutions," in M. Hechter, K.-D. Opp and R. Wippler (eds.) *Social Institutions: Their Emergence, Maintenance and Effects*, New York: Aldine de Gruyter.

Herring, R. (1990) "Resurrecting the Commons: Collective Action and Ecology," *Items* (SSRC) 44 (4): 64–8.

Huntington, S. (1993) "A Clash of Civilizations?" *Foreign Affairs* 72 (3): 22–49.

Kaplan, R. (1996) *The Ends of the Earth*, New York: Random House.

Karim, M.B. (1991) "Decentralization of Government in the Third World: A Fad or Panacea?" *International Studies Notes* 16 (2): 50–4.

Keohane, R.O. (1984) *After Hegemony: Cooperation and Discord in the World Political Economy*, Princeton: Princeton University Press.

Krasner, S.D. (ed.) (1983) *International Regimes*, Ithaca: Cornell University Press.

—— (1993) "Westphalia and All That," in J. Goldstein and R.O. Keohane (eds.) *Ideas and Foreign Policy*, Ithaca: Cornell University Press.

Krugman, P. (1994) *Peddling Prosperity: Economic Sense and Nonsense in the Age of Diminished Expectations*, New York: W.W. Norton.

Leatherman, J., Pagnucco, R. and Smith, J. (1994) "International Institutions and Transnational Social Movement Organizations: Transforming Sovereignty, Anarchy and Global Governance," Working Paper 5 (August): WP3, University of Notre Dame: Kroc Institute for International Peace Studies.

Libecap, G.D. (1989) *Contracting for Property Rights*, Cambridge: Cambridge University Press.

Lipschutz, R.D. (1991) "From Here to Eternity: Environmental Time Frames and National Decisionmaking," prepared for a panel on *"De-nationalizing" the State: The Transformation of Political Space, Social Time and National Sovereignty*, Vancouver: Conference of the International Studies Association.

—— (1992) "Reconstructing World Politics: The Emergence of Global Civil Society," *Millenium* 21 (3): 389–420; revised edn., J. Larkins and R. Fawn (eds.) *International Society after the Cold War*, London: Macmillan, 1996.

—— (ed.) (1995) *On Security*, New York: Columbia University Press.

—— (1997) "Networks of Knowledge and Practice: Global Civil Society and Environmental Protection Environment," in L. Anathea Brooks and S. VanDeveer (eds.) *Saving the Seas: Values, Scientists and International Governance,* College Park: Maryland Sea Grant College.

Lipschutz, R.D. with Mayer, J. (1996) *Global Civil Society and Global Environmental Governance: The Politics of Nature from Place to Planet*, Albany: SUNY Press.

Litfin, K.T. (1994) *Ozone Discourses: Science and Politics in Global Environmental Cooperation*, New York: Columbia University Press.

Long, N. (1992) Introduction to N. Long and A. Long (eds.) *Battlefields of Knowledge: The Interlocking of Theory and Practice in Social Research and Development*, London: Routledge.

McEvoy, A.F. (1986) *The Fisherman's Problem*, Cambridge: Cambridge University Press.

—— (1988) "Toward an Interactive Theory of Nature and Culture: Ecology, Production and Cognition in the California Fishing Industry," in D. Worster (ed.) *The Ends of the Earth*, Cambridge: Cambridge University Press.

Mayer, J. (1995) Personal communication, January 26.

Mearsheimer, J.J. (1995) "The False Promise of International Institutions," *International Security* 19 (3): 5–49.

Mitchell, D. (1993) "Public Housing in Single Industry Towns: Changing Landscapes of Paternalism," in J. Duncan and D. Ley (eds.) *Place/Culture/Representation*, London: Routledge.

Mitrany, D. (1966) *A Working Peace System*, Chicago: Quadrangle Books.

Nuijten, M. (1992) "Local Organization as Organizing Practices: Rethinking Rural Institutions," in N. Long and A. Long (eds.) *Battlefields of Knowledge: The Interlocking of Theory and Practice in Social Research and Development*, London: Routledge.

Onuf, N.G. (1989) *World of Our Making: Rules and Rule in Social Theory and International Relations*, Columbia: University of South Carolina Press.

Ostrom, E. (1990) *Governing the Commons: The Evolution of Institutions for Collective Action*, Cambridge: Cambridge University Press.

Ouchi, W.G. (1980) "Markets, Bureaucracies and Clans," *Administrative Science Quarterly* 25: 129–41.

Rosenau, J.N. (1992), "Governance, Order and Change in World Politics," in J.N. Rosenau and E.-O. Czempiel (eds.) *Governance without Government: Order and Change in World Politics*, Cambridge: Cambridge University Press.

Ruggie, J.G. (1983) "Continuity and Transformation in the World Polity: Toward a Neorealist Synthesis," *World Politics* 35 (2): 261–85.

—— (1989) "International Structure and International Transformation: Space, Time and Method," in E.-O. Czempiel and J.N. Rosenau (eds.) *Global Changes and Theoretical Challenges*, Lexington: Lexington Books.

Skocpol. T. (1985) "Bringing the State Back In: Strategies of Analysis in Current Research," in P.B. Evans, D. Reuschemeyer and T. Skocpol (eds.) *Bringing the State Back In*, Cambridge: Cambridge University Press.

Stone, D.A. (1988) *Policy Paradox and Political Reason*, Glenview: Scott, Foresman.

Thompson, J.E. (1992) "Explaining the Regulation of Transnational Practices: A State-Building Approach," in J.N. Rosenau and E.-O. Czempiel (eds.) *Governance without*

Government: Order and Change in World Politics, Cambridge: Cambridge University Press.

Tsing, A.L. (1993) *In the Realm of the Diamond Queen: Marginality in an Out-of-the-Way Place*, Princeton: Princeton University Press.

Wade, R. (1987) "The Management of Common Property Resources: Collective Action as an Alternative to Privatisation or State Regulation," *Cambridge Journal of Economics* 11 (2): 95–106.

Wæver, O. (1995) "Securitization and Desecuritization," in R.D. Lipschutz (ed.) *On Security*, New York: Columbia University Press.

Walton, J. (1992) *Western Times and Water Wars: State, Culture and Rebellion in California*, Berkeley: University of California Press.

Wapner, P. (1996) *Environmental Activism and World Civic Politics* Albany: SUNY Press.

Wendt, A. (1992) "Anarchy is what States Make of It," *International Organization* 46 (2): 391–425.

Wilmer, F. (1993) *The Indigenous Voice in World Politics*, Newbury Park: Sage.

Young, O.R. (1982) *Resource Regimes: Natural Resources and Social Institutions*, Berkeley: University of California Press.

—— (1994) *International Governance: Protecting the Environment in a Stateless Society*, Ithaca: Cornell University Press.

Young, O.R. and Osherenko, G. (eds.) (1993) *Polar Politics: Creating International Environmental Regimes*, Ithaca: Cornell University Press.

7   Toward a cosmopolitan bioregionalism

Mitchell Thomashow

The most daunting task facing the conceptual integrity of bioregionalism is its ability to convey metaphors, visions and practices that lend meaning to the complex interplay of local and global environmental relationships. This presents the inevitable, dual challenges of scale and meaning – how are personal and community actions relevant to the formidable complexities of global environmental change?

Bioregional theorists and activists are confounded if not disempowered by the conceptual challenges of interpreting dynamic global events and processes. How does a bioregional vision accommodate the bifurcation of economic globalization and political decentralization, the instability and dislocation of ecological and cultural diasporas, the elusiveness of pluralistic identities and multiple personas? How do people cultivate a meaningful and practical bioregional sensibility under such circumstances?

Bioregionalism emerges as a response to the formidable power relations of global political economy and the ensuing fragmentation of place. It seeks to integrate ecological and cultural affiliations within the framework of a place-based sensibility, derived from landscape, ecosystem, watershed, indigenous culture, local community knowledge, environmental history, climate and geography. More than an alternative framework for governance or a decentralized approach to political ecology, it represents a profound cultural vision, addressing moral, aesthetic and spiritual concerns. In effect, bioregionalism seeks to penetrate, inform and reinhabit the interstices of contemporary political economy, turning states and counties into biomes and watersheds, changing not only the boundaries of governance, but the boundaries of perception as well. Indeed, the reinhabitation of landscape is fundamentally a challenge of perception as well as citizenship.

Bioregional sensibility should develop ways of exploring spatial and temporal relationships that show the connections between place-based knowledge and global environmental change, the interdependence of local ecology and global economies, and the matrix of affiliations and networks that constitute ecological biodiversity and multicultural and multispecies tolerance – allowing different people to understand all the different places that may be considered home. This is the basis of a local/global dialectic and emphasizes the necessity of a cosmopolitan bioregionalism.

First, I will address two patterns that challenge bioregional sensibility: ecological and cultural diasporas and pluralistic identities. Here the tangible meets the abstract head-on through the leviathan of economic growth and global habitat degradation. Second, I will consider a perceptual reinhabitation of the landscape, grounded in specific suggestions regarding the ingredients of scale – space, place and time – that use a bioregional approach to carve a local/global dialectic.

Ecological and cultural diasporas

In stark contrast to the place-based rootedness espoused by bioregionalism is the deracination which accompanies habitat destruction. A habitat (by defini- tion) is typically a homeland for indigenous people and for a matrix of flora and fauna. When habitats are transformed by commercial, industrial and agricultural developments, natural resource extraction, tourism and war, a chain of ecolog- ical and cultural disruptions is initiated. Indigenous societies must either adapt to the changing circumstances, migrate to a new habitat, or face extinction. Although habitat transformation is intrinsic to global change, the accelerating pace of anthropogenic transformation which characterizes economic globaliza- tion has unleashed a process of extraordinary change that touches every corner of the globe.

What happens to people and species whose habitats are destroyed? In many cases, they face the silence of extinction. As Edward O. Wilson describes it:

> Extinction is the most obscure and local of all biological processes. We don't see the last butterfly of its species snatched from the air by a bird or the last orchid of a certain kind killed by the collapse of its supporting tree in some distant mountain forest. We hear that a certain animal or plant is on the edge, perhaps already gone.
>
> (1992: 255)

In the context of global economy, the same plight is endured by indigenous cultures whose local economy is no longer viable.

In many cases, people and species have one option – migration. Populations move between borders and habitats, seeking refuge, if only temporarily, until they can settle in a new homeland. The consequence of such migration is ecological and cultural upheaval. The problem of global refugees is increasingly severe. In his stunning and controversial book *The Ends of the Earth* (1996), Robert Kaplan vividly describes the human suffering which accompanies these upheavals. As indigenous cultures confront habitat destruction, they experience the dissipation of material life, and the possible dissolution of social stability and cultural integrity.

There is a correspondence between human refugees, escaped Ebola viruses and migrating songbirds whose winter and summer roosts have fallen to development. A habitat is transformed and new accommodations must be sought. We must consider the potential of widespread, global migrations of peoples and species as the shadow of globalization – returning to haunt the sheltered domicile of seemingly affluent domains, knocking on the doors of walled communities – a grim reminder of the transience engendered by forced uprootedness. Homelessness takes many shapes and forms.

Historically when migrating groups of people maintain the integrity of their culture despite their wanderings, we describe a diaspora, defined by Chaliand and Rageau as "the collective forced dispersion of a religious and/or ethnic group, precipitated by a disaster, often of a political nature" (1995: xiv). These groups maintain their cultural integrity via collective memory, which "transmits both the historical facts that precipitated the dispersion and a cultural heritage" (ibid.).

How do indigenous place-based societies retain their ecological knowledge when they no longer have access to their homeland? In the twenty-first century we face the prospect of multiple ecological and cultural diasporas, millions of migrants attempting to salvage their ecological and cultural integrity, using whatever means necessary to maintain some form of collective memory, or confront the dire consequences of naked assimilation and the loss of cultural identity. In the twenty-first century, having a homeland will represent a profound privilege. Living-in-place may become a quaint anachronism, reinhabitation a yuppie utopian vision.

Theories of island biogeography (Quammen 1996) and patch dynamics (Forman 1995) take on a new urgency. Ecologists and bioregionalists are developing proposals to deal with fragmented landscapes, considering how to "manage" viable, sustainable, integrated regions, given the mosaic-like nature of chopped habitats. Bioregionalists and conservation biologists propose ecological corridors, networks and regional dispersal strategies to integrate wild lands. Place-based communities increasingly resemble islands of diversity surrounded by homogeneous landscapes.

There remains an important challenge for bioregionalists: how are those rooted place-based communities to be allowed to become aware of their collective responsibility to lend support to those who are caught in diasporas? how is attention to be called to the magnitude of extinction? how are the scale of dislocation and the correspondence between threats to biodiversity, to cultural integrity and to human survival to be shown? No sense of place is complete without this awareness of the dynamics of cultural and ecological diasporas.

Ironically, as Aberley described in Chapter 2, bioregionalism emerged from the "baby boomer" diaspora of the postwar era. Certainly the "wanderings" of First Worlders are qualitatively a different form of uprootedness, but these people face uprootedness just the same, and they find bioregionalism so appealing because ecological and cultural integration is a prerequisite to sustainability and a basis for place-based identification. Given the prevalence and prospect of dislocation, identification with place, landscape, dwelling and habitat are rapidly becoming the locus of cultural stability and ecological continuity.

Bioregionalism projects a spirit of wholeness within community, a place-based foundation, grounded in the ecological nuances of the home territory. It cherishes diversity and pluralism without being overwhelmed by empires of commodity choices; it tolerates different ways of being, multiple formulations of identity without succumbing to a relentless, mindless quest to collect experiences. Bioregionalism presents an alternative to fragmentation by suggesting the construction of an ecological identity (Thomashow 1995), of orbits and connections that integrate mind and landscape, self and ecosystem, psyche and planet, without worrying about the paths not taken, but focusing instead on the tasks at hand – cultivating mindfulness about human/nature relationships in the service of both self-realization and community health.

Roger Kamenetz in *The Jew in the Lotus* (1994) writes about a remarkable encounter between an eclectic group of rabbis and the Dalai Lama. The Rabbis wonder why Tibetan Buddhism is so appealing and accessible to Jews. The Dalai Lama wonders how the Jews have been able to maintain their cultural and spiritual heritage in the midst of such a tumultuous diaspora. Although the Dalai Lama is speaking ostensibly for the future of Tibetan Buddhism, and is deeply concerned about how to maintain its integrity in the face of the occupation of its homeland, he is raising an issue that is increasingly relevant for all planetary citizens. By exploring how a place-based rootedness can be relevant to groups that temporarily lack places, bioregionalism has much to offer the displaced as well as the rooted.

Yet how important are particular places to human beings who are part of global economies? This chapter describes a cosmopolitan form of bioregionalism that breaks from the form of place-based bioregionalism found in Part I. I believe that bioregionalism should necessarily speak to the transient as well as the rooted, the sacred places on the land as well as the sacred places in the psyche, the cultural traditions of reinhabitation as well as the threatened knowledge of traditional place-based communities, the relationship between islands of diversity and chunks of homogeneous media.

Pluralistic identity and multiple personas

Whatever the cause of uprootedness, wherever its location, the consequential instability results in fragmented places and fragmented psyches. For those who have suffered from globalization, the first step toward cultural and ecological sustainability is to invent and construct viable forms of livelihood. Those societies that have the fortune of economic stability suffer from social discontinuity that is prompted by an array of choices and affiliations, without the groundedness of tradition and community. The material rewards for those societies that succeed in the global economy is based on a mere proliferation of consumer choices, the ability to have an extraordinary range of travel experiences, see countless entertainment events, read hundreds of books, cultivate dozens of relationships, move through many different virtual worlds, and participate in various organizations and institutions. Multiple personas are created in the multiple worlds of global economy. The consumers of global economy shift allegiances but have no true affiliation. Nowhere is this more evident than on the internet where one can enter a "chat group" and take on a fantasy persona.

Global economy provides the prospect of constructing a matrix of personal identities which can be chosen from one's ethnicity, sexuality, gender, expertise, political ideology, among other possibilities. Personal identity reflects a revolving array of interests and affiliations. This is the challenge of constructing a pluralistic identity in the global economy, of the multiple personas available to the psyche: we can invent ourselves, but where is the center?

Nevertheless, it takes a grounded person to negotiate these competing realms of identity and difference. As consumers, we face what David Gergen (1991) refers to as "saturation" – the feeling of being overwhelmed by an array of identity choices. Beneath the surface of the saturated person lie the more threatening perceptions of discord and fragmentation that permeate the layers of multiple personas and the consumer's relentless search for foundation and grounding. A bioregionalist must address, understand, and overcome the perceptions of discord that accompany economic globalization, and ecological and cultural diasporas. These processes dramatically impact the psyche, challenging personal and communal identity. No person or community is immune from these trends. Rather, the search for identity reflects an emerging human psyche which dwells in a world-in-flux. Contemporary bioregionalism necessarily includes an ecopsychology of global change – a place oriented agenda for everyday decisions, grounded in material life, cultural exchange and ecological relationships.

The perceptual foundations of a bioregional sensibility

At the heart of a bioregional sensibility is the concept of place-based reinhabitation. To engage in reinhabitory practice is to challenge the human imagination. Much work has been done to develop a contemporary place-based knowledge using concepts such as ecological literacy (Orr 1992), ecological identity (Thomashow 1995), deep ecology (Naess and Rothenberg 1989), mapping (Aberley 1993) and literary exploration (Tallmadge 1997). Place-based sensibility remains a vital theme among American nature writers – merging imagination with natural history, autobiography with local ecology. This theme emerges frequently and beautifully in the works of Terry Tempest Williams (1991), Gary Snyder (1990), Barry Lopez (1978), Scott Russell Sanders (1993), Ann Zwinger (1989), Pattiann Rogers (1994) and Robert Pyle (1993), among hundreds of other superb writers and thinkers. The spirit of bioregionalism is thoroughly guided by these visions.

Place-based knowledge is meaningful not only as a commitment to understand local ecology and human relationships but as a foundation from which to explore the relationships between and among places. To explore a local place is to engage in the rich textures of microhabitats, the tapestry of stories regarding how human populations settled in those habitats, stories of exploitation and restoration, degradation and reinhabitation.

There is a tangible quality to the explorations of place. Consider Thoreau's comments on what is a reasonable boundary of exploration:

> My vicinity affords many good walks; and though for so many years I have walked almost every day, and sometimes for several days together, I have not yet exhausted them. An absolutely new prospect is a great happiness, and I can still get this any afternoon. Two or three hours' walking will carry me to as strange a country as I expect ever to see. A single farmhouse which I had not seen before is sometimes as good as the dominions of the King of Dahomey. There is in fact a sort of radius discoverable between the capabilities of the landscape within a circle of ten miles' radius, or the limits of an afternoon walk, and the threescore years and ten of human life. It will never become quite familiar to you.
>
> (Thoreau 1975: 598)

Thoreau suggests that there is an appropriate scale for exploration.

Global economy requires that bioregionalists explore both the immediate landscape (place) and those larger systems that exist beyond the horizon (space). The local landscape can no longer be understood without reference to the larger patterns of ecosystems, economies and bureaucracies. Gary Snyder maintains that "to know the spirit of a place is to realize that you are a part of a part and that the whole is made of parts, each of which is whole. You start with the part you are whole in" (1990: 38). A bioregionalist sensibility moves from the parts to the whole, but in a way that lends meaning to both, and is tangible to the psyche. What types of "explorations" and experiences contribute to the realization of both place and space?

As Thoreau suggests, you can become increasingly intimate with the nuances of the landscape, but it will never be totally familiar to you. There are just too many patterns to observe. With patience, perseverance and attention, one can recognize patterns of change – seasonal variations, where the vernal photosynthetics make their first appearance, when the barred owl is most likely to call, which years have the most prolific acorn crops.

One's perceptual field of vision can evolve to sense the important changes in the places we inhabit and the spaces we occupy. Jacob Von Uexhill invented the term *umwelt* to refer to the perceptual environment of a biological organism. Umwelt refers to what an organism can perceive, given the organism's biological make-up. Fraser suggests that there are temporal features of sense-specific umwelts, writing that: "For each animal the world-as-perceived is determined by the potential functions of the totality of its receptors and effectors" (1975: 75). By taking the notion of umwelt seriously, we understand that "creatures of different psychobiological organization (and in the same creature, different senses) might eventually be understood as grasping reality differently" (ibid.). A place-based bioregional sensibility recognizes features of specific umwelts. Our organismic umwelt is very much bound by scale.

David Rothenberg shows that technology represents the amplification of umwelt, changing how we perceive nature, but also reminding us that it "always remains beyond us, something to wish for, still far away, just out of reach" (1993: xvi). By looking through a telescope, one can change the scale of one's vision, making distant objects seem close, shrinking the expanse of time and space. By looking through a microscope, one can get much closer to the minuscule, noticing how much space there is in even the smallest distances.

How can the expansion and amplification of umwelt allow the bioregionalist to integrate the local and the global? Interpreting global patterns requires the expansion of umwelt so that it encompasses more than the places we live in. One must move beyond that which can be directly perceived to consider a full range of spatial and temporal dimensions.

I start with a garden plot. I have worked this soil for seventeen years, cultivating this small portion of a forested landscape. If I leave the plot untended for any length of time, it is reclaimed by the forest, starting with the virulent, thorny blackberries and the ubiquitous oak seedlings sprouting from last year's acorns. As I work the soil, I plunge my spade into countless pebbles and small boulders. To understand where they come from I must step out of my organismic umwelt and incorporate the conceptualization of a larger time frame. These rocks were deposited when the landscape was covered with ice during the last glacial interval 10,000 years ago – not so many years in geological terms, but many more years than I can readily perceive. How much harder it is to travel 10,000 years into the future and imagine what this landscape might look like. From a spatial perspective, I notice how the garden is an edge, a demarcation that separates a space in the Northern forest. As such it represents a different habitat for plants and animals. Can I use my imagination to explore the umwelt of those creatures that find homes in the garden?

I watch the stream as it rises and falls with the rains and snows, running fast in April, slowing to a trickle in August. It is easy enough to observe the relationship between rainfall, snow melt and stream flow. Can I follow the stream all the way to the river and the sea?

To answer these questions, one must carefully note the importance of bioregional history, expand and amplify one's organismic umwelt, and juxtapose space with time (i.e. space-time). The ability to juxtapose scale requires discipline, the necessity of clarifying boundaries, and, somewhat paradoxically, crossing the boundaries of space and time. In *Toward a Unified Ecology* (1992), Timothy Allen and Thomas Hoekstra emphasize the importance of space-time as a means of interpreting complex natural systems and ecological patterns. Ecological processes are multiscaled. Ecological interpretation requires the use of multiple criteria – landscapes, ecosystems, communities, organisms, populations, biomes and biospheres. With respect to the juxtaposition of space and time, Allen and Hoekstra write:

For any level of aggregation, it is necessary to look both to larger scales to understand the context and to smaller scales to understand mechanism; anything else would be incomplete. For an adequate understanding leading to robust prediction it is necessary to consider three levels at once: (1) the level in question; (2) the level below that gives mechanisms; and (3) the level above that gives context, role, or significance.

(Allen and Hoekstra 1992: 9)

This observation has extraordinary theoretical and practical power; it encompasses a perceptual wisdom that is often neglected and has been forgotten. It provides wonderful guidance for how a bioregional sensibility might approach global environmental change.

For example, I live in the dry oak–beech–maple, Eastern deciduous Northern woodlands. Through familiarity and study, I recognize the most prevalent flora and fauna of the habitat, although I have virtually no knowledge of the soil microorganisms. To completely and thoroughly interpret the "level in question" I should be able to identify, study and observe most of the plants that live in a chosen plot of woodland. This will allow me to notice subtle changes and nuances, especially those that occur within measurable and perceivable time frames. This is true regardless of the methodology I employ. I can use quantitative yardsticks or I can rely on intense qualitative attention, what we might distinguish as scientific observation and vernacular wisdom.

In either case, my interpretive power is limited to specific time and space boundaries which are constrained by two qualities: my perceptual limits (what I can see or measure) and my conceptual limits (the extent to which I incorporate multiple levels in my thinking). My understanding of mechanism will be greatly enhanced by attending to microscopic scales, which may include soil chemistry, microbial biology, plant physiography and other sets of inquiry. My understanding of context requires attention to biogeography, climatology, geomorphology and so on. I don't wish to limit these categories to the conceptual frameworks of contemporary science. Extraordinary insights occur when we perceive the relationships between boundaries.

If I observe drought conditions – a lack of rainfall, water-starved plants, or diminished bioproductivity, my insights are incomplete without linking these observations to additional interpretive levels. Why is there no rain? To answer this, I observe atmospheric circulation which requires weather maps and perhaps an understanding of paleoclimatology. What patterns will these levels help me detect? Or perhaps I must rely on the deepest levels of intuition and awareness, finding patterns in the movement of the clouds, wind direction, changes in how the air tastes, smells and even weighs. How do plants and animals respond to this lack of rainfall? What pathways help me understand the cellular level of response?

People often ignore or deny the implications of multiple, interacting levels. Illness is more than discomfort and dysfunction, and we cannot explain it by attending to microbes alone, or by solely exploring the context (organismic stress). To become well, we must understand and cure the illness at multiple levels. This is the conceptual foundation of holistic health. We cannot solve the crime problem merely by providing law-abiding citizens with guns or by building more prisons. Such solutions rely exclusively on narrow observations of the level in question. However, many social, cultural and environmental problems are perceived from extremely limited frameworks.

Bioregional initiatives inevitably encounter the complexities of scale and meaning. An appealing compromise may be reached locally, only to have it subverted by global capital (Gould, Schnaiberg and Weinberg 1996) or by the state (Lipschutz 1996). This can be devastating for activists who are left wondering whether their actions will really ever change anything. Activists should be savvy to the scale of their actions. On one level, it may appear that an initiative has failed, but on another level, an entirely different process may be occurring. As Lipschutz described in the last chapter, the emergence of global civil society reflects a domain that exists between boundaries and home places, a synergy of influence that transcends place. With respect to global civil society, bioregional activism may have a positive impact beyond the boundaries of a particular home place (i.e. may have influence on the space between places).

When people search for their roots, they recognize the depth of their uprootedness. They discover that their affiliations are broad and vast, not necessarily linked to any specific place, but rather a constellation of places. The delineation of hard and fast boundaries is the cause of much human suffering, as clashing tribes or nation-states argue about who belongs where. Bioregionalism must avoid the shadow of extreme regional identification. Rather, strong communities allow for permeable boundaries, and recognize the connections between places as intrinsic to the well-being of any one place.

In global economy, people identify with many places at once, forming networks and allegiances based on pluralistic identities. This is the essence of a local/global dialectic in which regions unfold within and between each other. A bioregional sensibility must also cultivate a language for expressing the connections between regions. People and commodities from other places "wander through" a bioregion. A bioregion is the stopping place for the migration of assorted flora and fauna, each of which makes its indelible imprint on the ecology and culture of the neighborhoods where they temporarily reside. A myrtle warbler has two addresses. The white winged cross-bill floats through numerous boreal habitats in search of bountiful pine cones on spruce trees. Migrating species have much to offer regarding what it's like to live in different places. There will always be trade and exchange between places. The cultural and ecological relationships that exist between places are transregional.

Ideas and concepts move in and out of a particular place, not attached to a region but expressed through an array of mediums, vectors, or paths – the landscape, air, water, the spoken word and modern technology. These are "mind regions" that cut through bioregional distinctions; metaregions where the exchange of ideas percolate through consciousness as the wind sweeps through the trees on a gusty day. Ideas move quickly and change with each iteration. But these ideas often settle in a place. They congeal and coagulate, attract supporters and detractors, and form world-views and paradigms that attach temporary order to a complex universe. Metaregional affiliations transcend local knowledge and inform the world of these ideas. A bioregional sensibility provides a basis for those ideas, by sorting them through a local sieve, a place-based gristmill, and considering whether they have any insight or explanation to offer.

There is also the integrated global circulatory system of the world. These are the patterns of biogeochemical cycles, the currents of the oceans and winds, the great weather systems that pass through every bioregion, the continental plates that drift beneath your feet, the complex interplay between microbial life and atmospheric circulation, or what is referred to as Gaia. These are the systems of global environmental change, where spatial and temporal variation are least accessible, but crucially important.

Cosmopolitan bioregionalism for perceiving global change

Developing the observational skills to patiently observe bioregional history, the conceptual skills to juxtapose scales, the imaginative faculties to play with multiple landscapes, and the compassion to empathize with local and global neighbors – these qualities are the foundation of a bioregional sensibility. The challenge for the bioregionalist is to learn how to see the world through such a lens and then to bring those insights, patiently and strategically, to classrooms, public forums, legislatures, media networks, nature centers, or wherever people congregate to learn about their lives and make decisions about their future. How can the bioregionalist engage communities in thinking about global change?

A bioregional sensibility requires multiple voices and interpretations. There are many paths to bioregionalism, and its perceptual vision is likely to reflect the diverse experiences of its practitioners. In the spirit of such diversity, in respect for the experimental and grass-roots quality of bioregional efforts, in honor of the intricacies and complexities of a local/global dialectic, a bioregional sensibility is necessarily open-ended and flexible.

To conclude this essay, I describe perceptual guidelines that link the bioregional with the global, in order to construct a cosmopolitan bioregionalism:

Study the language of the birds Integrate language and landscape. Make the study of flora, fauna, landscape and weather a daily practice. Know what species coinhabit a community. Know who is just passing through and where they are going. Learn from the ecosystem. Tell stories about wildlife and landscape as a means of revitalizing the spirit and psyche, of honoring the diversity of species, of expanding the notion of community. Restore natural history to collective memory so that it is no longer endangered knowledge.

Navigate the foggy, fractal coastline Understand that different scales may yield contrasting observations and that different people will have various interpretations. Avoid the illusion of contrived stability. Local knowledge requires practitioner-based science and place-based wisdom, cadres of bioregional investigators who catalog the dynamics of local environmental change in their home communities, who compare notes with their colleagues, who can chart a steady course in the midst of complex, turbulent change.

Move within and without Trace the ecological/economic pathway of everyday commodities to fully understand the impact of globalization – its benefits and threats. Consider the full matrix of citizenship, all of the ways that speech, intentions, motivations and actions contribute to the formation of a bioregional sensibility.

Cultivate a garden of metaphors Pay attention to sensory impressions and their broader symbolic meaning. Find the metaphors of anxiety that illuminate the relationship between the psyche and the planet. Find the metaphors of wholeness that pervade good nature writing – fruitful darkness, turtle island, attentive heart, crossing open ground, the spell of the sensuous, the island within – and contemplate their meaning. Trace the ecology of imagination.

Honor diversity Use different ways of thinking and various cultural perspectives as a conceptual lens. Understand the world through the eyes, ears and nose of wild creatures. Incorporate multiple learning styles. Attend to difference by exploring what is common and learning from what remains different.

Practice the wild Experience wild nature and wild psyche. Consider the stark reality of the food chain. Observe how civilization can never keep the wild completely at bay. Let wild nature inform play, work, love and worship. Practice the wild to balance the civilized.

Alleviate global suffering Have compassion for the chasms of despair. Find the holes in the bioregion, the places of darkness that require healing and attention. Understand how the fruits of affluence often hinge on the exploitation of the weak. See the world as it is, without blinders, transcending denial.

Experience planetary exuberance Life bursts forth everywhere. It is an indomitable, ever-present, mysterious force that permeates every surface of the biosphere, every pore in your skin. Every life-form is a unique expression of the poetic and sublime.

Above all else, bioregionalism strives to allow communities to celebrate their psyches, habitats and ecosystems. Through the efforts and experiments of bioregional governance, the perceptual groundwork is being prepared for a cosmopolitan bioregionalism. But this will only occur in the spirit of developing interactive, participatory learning communities, with spaces for contemplation, and paths of exploration. In order to achieve a frame of mind that acknowledges the magnitude of global and personal change, cosmopolitan bioregionalism represents a way of integrating psyche and nature for the purpose of constructing meaning and interpreting the world.

Bibliography

Aberley, D. (ed.) (1993) *Boundaries of Home: Mapping for Local Empowerment*, Gabriola Island: New Society Publishers.
Allen, T.F.H. and Hoekstra, T.W. (1992) *Toward a Unified Ecology*, New York: Columbia University Press.
Barber, B. (1995) *Jihad vs. McWorld*, New York: Random House.
Barnet, R.J. and Cavanagh, J. (1994) *Global Dreams*, New York: Simon & Schuster.
Chaliand, G. and Rageau, J. (1995) *The Penguin Atlas of Diasporas*, New York: Viking.
Forman, R.T.T. (1995) *Land Mosaics*, New York: Cambridge University Press.
Fraser, J.T. (1975) *Of Time, Passion and Knowledge*, New York: George Braziller.
Gergen, D. (1991) *The Saturated Self*, New York: Basic Books.
Gould, K.A., Schnaiberg, A. and Weinberg, A.S. (1996) *Local Environmental Struggles*, New York: Cambridge University Press.
Kamenetz, R. (1994) *The Jew in the Lotus*, San Francisco: Harper.
Kaplan, R. (1996) *The Ends of the Earth*, New York: Random House.
Lipschutz, R.D. with Mayer, J. (1996) *Global Civil Society and Global Environmental Governance*, Albany: State University of New York Press.
Lopez, B. (1978) *Crossing Open Ground*, New York: Random House.
Mander, J. and Goldsmith, E. (eds.) (1996) *The Case Against the Global Economy*, San Francisco: Sierra Club Books.
Naess, A. and Rothenberg, D. (1989) *Ecology, Community and Lifestyle*, New York: Cambridge University Press.
Orr, D.W. (1992) *Ecological Literacy: Education and the Transition to a Postmodern World*, Albany: State University of New York Press.
Pyle, R. (1993) *The Thunder Tree*, Boston: Houghton Mifflin.
Quammen, D. (1996) *The Song of the Dodo*, New York: Scribner.
Rogers, P. (1994) *Firekeeper*, Minneapolis: Milkweed.
Rothenberg, D. (1993) *Hand's End*, Berkeley: University of California Press.
Sanders, S.R. (1993) *Staying Put*, Boston: Beacon Press.
Snyder, G. (1990) *The Practice of the Wild*, San Francisco: North Point.
Tallmadge, J. (1997) *Meeting the Tree of Life: A Teacher's Path*, Salt Lake City: University of Utah Press.
Thomashow, M. (1995) *Ecological Identity*, Cambridge: MIT Press.
Thoreau, H.D. (1975) "Walking," in C. Bode (ed.) *The Portable Thoreau*, New York: Viking Press.
Williams, T.T. (1991) *Refuge*, New York: Random House, 1991.
Wilson, E.O. (1992) *The Diversity of Life*, Cambridge: Harvard University Press.
Zwinger, A. (1989) *The Mysterious Lands*, New York: Penguin Books.

8 Climate-change policy from a bioregional perspective
Reconciling spatial scale with human and ecological impact

David L. Feldman and Catherine A. Wilt

One of the principles that animated the United Nations Conference on Environment and Development (UNCED) Local Agenda 21 has been the implicit belief that a bioregional-type scheme for managing climate-change and other large-scale global environmental problems (e.g. biodiversity) has considerable merit. While this approach has not always been articulated consistently in Agenda 21, it is nonetheless apparent in this program in three ways: (1) by its emphasis on encouraging the restructuring or adjustment of decision-making to integrate socioeconomic and ecological considerations within the same sets of policies; (2) by encouraging the de-bureaucratization of policy-making through its emphasis on widescale localized participation in critical development and environment decisions; and (3) by an emphasis on integrated watershed and other resource-shed development to forestall erosion and deforestation, to encourage wise land-use, and to ensure the participation of indigenous peoples in environmental decisions (Grubb *et al.* 1993). To some extent, however, this same Local Agenda 21 initiative also exemplifies some of the difficulties inherent in applying a bioregional approach to the management of climate-change. Two problems are paramount: (1) identifying a decision-making "locus of control" (who, ultimately, is responsible for policy formulation and implementation under global climate-change agreements?); and (2) assessing and managing climate-change impacts (how, in the final analysis, do we formulate viable regional responses to a problem that has ubiquitous spatial and temporal causes and consequences?).

This chapter first examines the relevance of bioregionalism and its focus upon exploiting the use of "natural regions" as potentially important tools for furthering our understanding of the causes and consequences of anthropogenic global climate-change.[1] Second, we evaluate whether decision-makers may better manage needed institutional and societal responses to climate-change through adopting bioregional approaches that emphasize the adoption of a "land ethic" and associated perspectives (McGinnis 1993). In evaluating the utility of a bioregional approach to climate-change management, we consider two questions: First, are there "real world" bioregionally based-approaches for managing climate-change that have been adopted by decision-makers? If so, how do they function and what have been their results? Second, in the absence of bioregional approaches in current use, can we conceive of ways of incorporating a bioregional approach into the management of climate-change (e.g. an integrated human institutions–resources based approach)? We contend that a bioregional approach may help better articulate policy responses to climate-change. However, it has yet to be proven that a bioregional approach in and of itself adds significant value to existing management frameworks or provides durable, efficacious mechanisms appropriate to this problem. This challenge needs to be further examined.

Some bioregional characteristics of the climate-change problem

Many activities that contribute to global climate-change (e.g. energy production/consumption, deforestation and transportation choice) are influenced by, and occur within, activity sectors in individual states, provinces, regional districts, or even cities and metropolitan regions within countries. These activities have multiple, transboundary and transgenerational effects. For example, the coal currently being burned in power plants in the Tennessee Valley generates carbon dioxide gas that will reside in the atmosphere for decades, contributing its radiative forcing potential to global warming and, eventually, to sea-level rise affecting Bangladesh, Egypt and Vanuatu (the former New Hebrides). Likewise, demand for hardwoods for furniture in Japan, for example, or to develop orange groves for US or European consumers may lead to pressures for the destruction of rain forests in Indonesia, Brazil, or Honduras. This, in turn, may lead to the diminution of a large "carbon sink" able to absorb trace gases such as carbon dioxide and thus abate global warming. In short, such activities meld natural and social connections in complex, unanticipated ways.

What these examples point to is one often-overlooked fact about climate-change – the conditions affected by global warming (e.g. agriculture, coastal zone and water resources management), and the activities that produce them – must both be mitigated subnationally. For this reason, bioregionalism is potentially relevant to the management of climate-change for three reasons:

1   Given the immensity of the possible impacts of climate-change, and their varied scale, the effective management of global climate-change will require dedicated subnational responses within countries (Feldman and Wilt 1996). In partial recognition of this problem, UNCED and Agenda 21 encourage partnerships between different levels of government and between governmental and non-governmental organizations (NGOs) in order to hasten sustainable responses to these threats (Frost 1994: 169). Moreover, the Global Environment Facility – a cooperative effort of the United Nations Development Program (UNDP), United Nations Environment Program (UNEP) and the World Bank – funds adaptation projects for climate-change in several subnational regions in many developing countries.

2   There has been a dramatic rise in subnational (e.g. "state"-, provincial- and regional-level) global change related activities in both developed and less-developed nations since the late 1980s (Feldman and Wilt 1994; Wells 1993; Jones 1991; Grubb *et al.* 1993). Bioregionalism may provide one means of evaluating – or holding up to some set of standards – the efficacy of such policies. These policies include energy efficiency and conservation, transport planning and regional impact-assessment programs. In some instances, these activities have been codified in the climate-change "action plans" of so-called "Annex I" signatories (e.g. highly industrialized nations) to the *United Nations Framework Convention on Climate Change*. These countries are obligated, under this convention, to stabilize their emissions at 1990 levels by the year 2000.

There have also been numerous activities at the subnational level in developing nations (that are not required to reduce greenhouse gas emissions by the Framework Convention). These activities have been encouraged, in many instances, by NGOs operating at district- and even village-level in nations such as Thailand, Indonesia and Malaysia (Sadduzzaman 1994; Chiang 1994; Sari 1994).

3   A number of scholars suggest that countries must aggressively develop subnational strategies to reduce greenhouse gases and – perhaps more importantly – adapt to the impact of climate-change (Devall and Parresol 1994; Torrie 1993; Feldman and Wilt 1996). This is significant because, by many accounts, climate-change is inevitable and the best societies can now hope to achieve is some measure of adaptation to its impact while simultaneously decreasing their anthropogenic emissions of carbon dioxide, methane and other so-called "greenhouse gases" to forestall even worse consequences (Watson, Zinyowera and Moss 1996). Many scientists believe that a discernible human influence on global climate has already occurred due to the past century-and-a-half's industrial activities. The residence time in the atmosphere of most radiative forcing gases (e.g. carbon dioxide and methane); the temperature change that already may have occurred or that is projected to soon occur; and the likely near-term trends in greenhouse gas emissions under even the most conservative, energy-conserving scenarios, lead to a forecasted global average warming of between 1 and 3.5 degrees Celsius by 2100 (Watson, Zinyowera and Moss 1996: 6).

Reasons given for advancing subnational policies in these studies include many that resonate with bioregional perspectives. In most instances, these reasons fall into two general categories: functional and cultural. The former include the fact that environmental degradation, including that anticipated to occur from global climate-change, ultimately depends upon local geographical and biological conditions and resources, e.g. the location and amount of available water, the type of vegetation and the physical constraints on the ecosystem in a given region, such as physical proximity to, or isolation from other threatened (or even nonthreatened) ecosystems, and the ability of flora and fauna to adapt to environmental change. The latter includes the need for flexible, decentralized programs that are publicly acceptable, that directly involve affected stakeholders in their formulation and implementation, that are comprehensive (i.e. sensitive to the synergies among resources and institutions), innovative, globally accountable, and that acknowledge the unique evolution of political economies and histories within distinct regions (see Chapter 6 in this volume by Lipschutz; Torrie 1993; Devall and Parresol 1994; Berlin Communiqué 1995; US Congress, OTA 1993a; Dryzek 1987; Mann 1990).

Given both sets of reasons, bioregions may be essential to understand the barriers to – and the means to facilitate – global governance of climate-change because they constitute the areas within which the causes of global change (e.g. energy consumption, deforestation) are generated. Bioregions are the places where the most serious impacts of this change (e.g. sea-level rise, threats to food supplies) are actually felt. As Lipschutz and others point out, it is impossible to define the array of human interactions that produce large-scale global environmental changes, or those that result from them, without understanding that the interaction of natural, political, economic and cultural factors that lead to both are the result of settlement patterns, economic activities and values toward the environment that are manifested within highly localized settings, not "nations" (Lipschutz, Chapter 6 in this volume; Feldman and Wilt 1996). As a result of these claims, bioregionalism may be especially relevant to global climate-change given the spatial uncertainty of its effects, as discussed below.

Natural and social system relevance and constraints

By most accounts, the regional impact of climate-change is thought to be highly variable and subject to "surprises," nonlinearities and uncertainties. Uncertainty in the natural systems affected by climate-change is compounded by problems of geographic indeterminacy in General Circulation Models (GCMs) used by atmospheric scientists to predict specific regional consequences on rainfall, temperature, cloud cover and other climatological factors. In short, GCMs are simply too inexact to be able to resolve the spatial impacts of climate-change into regions small enough for traditional administrative planning, mitigation and adaptation activity (NGA 1990; Morandi 1992).

While the average impacts of climate-change are predicted to be relatively mild on a global scale, with temperatures increasing perhaps 1–3.5 degrees Celsius over the next 100 years (Nordhaus 1991; Watson, Zinyowera and Moss 1996), its positive and negative effect on social systems may be quite strong within specific regions (e.g. the effect of sea-level rise on coastal zones, or protracted drought in farming regions). A large country such as the US has a sufficiently diverse portfolio of regions and economic activities to ensure that the average impact of climate-change remains quite small. However, the risk of significant environment impact is much higher for less-diversified economies within less-developed nations.

The inability of GCMs to resolve spatial impacts into areas suitable for regional or other forms of subnational planning makes the assessment of the consequences of climate-change difficult. Thus the creation of public policy that can address those consequences becomes harder too. In theory, while this spatial uncertainty constitutes a potential opportunity for adopting a bioregional approach to climate-change, in practice, it suggests at least four important barriers to policy change:

1 Because global climate-change is an uncertain, potentially long-delayed concern, there is limited incentive for individuals and institutions – particularly in smaller, subnational regions jurisdictionally required to perform a broad array of functions (e.g. public safety, resource management) – to suddenly react to this uncertain threat by engaging in significant economic sacrifices (McLaren and Skinner 1987).

2 The experience of other environmental areas suggests that uncertainty is likely to force decision-makers to view the problem as remote, and not a cause for concern within the span of a normal term of office or other policy horizon. They are unlikely to change conventional institutional forms of management or to convert these doubts into a clarion call for action, since they cannot know the real costs that might be incurred in doing so (Lee 1993).

3 Contradictory information about the natural and social worlds makes a bioregional approach untenable in the eyes of some policy-makers. Energy, environmental, natural resource, land-use, transportation, regional planning and waste management policies may have an incidental impact on climate-change, but are not intended as global change policies. While the benefits of a comprehensive, integrated "bioregional" approach to climate-change may in the long run prove beneficial, in the shorter run, there is unlikely to be a constituency as strong in its support for such a comprehensive policy as there is for these separate policy areas.

4 Finally, there is the so-called "collective action problem." Local or regional jurisdictions may have little incentive to engage in climate-change mitigation and abatement. The ozone layer and global climate are obvious "public goods." Everyone has a stake in the benefits they provide, so everyone acts as a free rider and, therefore, does not contribute to the mitigation of the problem. It is unclear why an individual should undertake local activities that potentially benefit others at one's own expense. This is a classic "free rider" problem (Wilt and Feldman 1995).

The next section considers what some of the motives for a comprehensive, integrated approach might be.

The sustainability conundrum

Many of the major policy initiatives pertinent to the management of climate-change, and which have recently been taken up by nations working together, are based on an assumption that bottom-up grass-roots approaches that integrate natural and cultural systems within small regions are effective strategies to deal with the many issues and concerns endemic to climate-change. Furthermore, these policy initiatives predicate that traditional bureaucratically defined spaces are inconsistent with ecological and cultural resource needs. These initiatives include the aforementioned *Framework Convention on Climate Change*, a variety of international policy declarations, and several institution-building efforts (such as Agenda 21). In this section, we argue that attempts to implement many of these initiatives resonate with bioregional ideals. They also suggest some avenues for serious implementation of a bioregional approach – and some important challenges to consider.

International organization activities

Since 1992, most efforts to establish subnational climate-change initiatives have taken as their point of departure the desire to formulate local, regional and provincial sustainable development strategies. The concept of sustainable development was introduced in the early 1980s in the World Conservation Strategy conjointly formulated by World Wildlife Fund (WWF), the International Union on the Conservation of Nature (IUCN) and UNEP. It was later reiterated by the 1987 Brundtland Commission (Furuseth and Cocklin 1995: 244). Specific locally or regionally focused strategies were inspired by the United Nations Conference on Environment and Development (at the Earth Summit) and its promulgation of Agenda 21 – a blueprint for regional sustainability.

While the concept of sustainable development does not command universal agreement, in the context of a bioregional framework for managing global change, it has introduced some common principles into the debate. First, and of particular relevance to the climate-change debate, most proponents of sustainable development agree that a sustainable policy approach is one that steers a compromise between unlimited growth and unlimited protectionism on the one hand, and between strong government controls on population and economic growth and unfettered markets on the other (Furuseth and Cocklin 1995: 245; Haas 1996). Institutionally, in other words, the concept of sustainable development calls into question bureaucratic institutions for the management of natural resources that are predicated largely on the principle of unlimited growth and the infinite malleability of resources.

Second, it also suggests – and this makes it particularly relevant for climate-change (as well as difficult to operationalize) – that environmental protection, development, poverty and democratization cannot be "unbundled." Solving one problem requires addressing all of them together, in an integrated fashion (Haas 1996). Certainly, proponents of a proactive climate-change policy have long insisted that an efficacious climate policy must seek to reduce energy use, encourage energy conservation, discourage deforestation through creating regionally sustainable alternative economic activities, and reduce population growth (Watson, Zinyowera and Moss 1996).

The 1992 *UN Framework Convention on Climate Change* requires that all signatories develop and issue national inventories of greenhouse gas emissions and sinks. It also requires developed country signatories to undertake greenhouse gas assessment and mitigation programs and to develop specific regional resource management plans (e.g. integrated coastal zone programs, river- and lake-basin protection programs, agricultural resource-management programs and protection of sensitive areas most vulnerable to drought and desertification). Several developed or "Annex I" countries under the *Framework Convention on Climate Change* – most notably the United Kingdom and New Zealand – have taken both of these principles seriously enough to try and develop regionally based sustainable development plans that are designed to address climate-change among an array of other issues. In both nations, local and regional input is encouraged in the environmental and natural resource planning process – in part, to help better manage the possible impacts of climate-change (Richards and Biddick 1994; Furuseth and Cocklin 1995; Morphet and Hams 1994).

In the UK, for example, many "local authorities" (the basic units of local government) have chosen to develop such plans on their own initiative. What is significant about these self-initiated plans is that these local authorities have established a Local Government Management Body (LGMB) in an effort to promote and synthesize best-management practices across jurisdictions and to ensure uniform national objectives for environmental priorities, particularly in such areas as transport, purchasing and economic development (Morphet and Hams 1994: 480). In the US, while the Environmental Protection Agency provides guidance and input into the process of establishing priorities, states and local communities choose to follow their own approaches. Thus, one possible lesson here is that states might be encouraged to develop their own uniform objectives through state-initiated process, perhaps through the Environmental Commission of the States, or by another organization.

Finally, as in the US, these foreign efforts seek to empower local citizen organizations to articulate priorities and other issues, and to take direct action to improve the rural environment (Morphet and Hams 1994; Martin 1995). For example, in the UK, local sustainable development plans encourage grassroots participation through a Community Action Experiment Program that empowers local groups to better understand and articulate issues and to take direct action to improve local environmental conditions.

Similar bioregional-type efforts are taking place in New Zealand and Thailand (Atkinson and Vorratnchaiphan 1994; Furuseth and Cocklin 1995). In the latter, several women's and environmental groups and labor unions have helped formulate environmental plans by "Local Environmental Action Committees." These committees are designed to assist local governments (who have few staff and other resources) in providing technical advice and program evaluation guidance. In short, these activities appear to be based on the desire to address both public choice and environmental justice concerns. The former is encouraged through involving grass-roots stakeholders directly in regionally based decision-making activities, and not just polling them on their preferences; the latter is promoted through ensuring distributive fairness (i.e. by incorporating minority participation and gender equity in these activities).

How durable and successful these efforts are remains to be seen. Like the efforts taking place in American states and communities, these foreign efforts are in relatively early stages of development and implementation. However, as US subnational priority-setting efforts continue, these foreign efforts provide examples of regional priority-setting that are worth monitoring.

Other international activities

International calls to action prescribe bioregional-like policy responses to climate-change. Agenda 21, a set of principles to encourage sustainable development to be implemented through the newly established Commission on Sustainable Development, promotes international funding of national greenhouse gas abatement projects and programs based on a bottom-up approach to development planning. It is supposed to encourage information exchange between local communities, regions and national governments in order to encourage environmentally appropriate economic practices including small-scale energy development, afforestation and conservation.

According to Agenda 21, integrated sustainable development programs designed to reduce greenhouse gas emissions should be implemented at local and regional levels where "open governance responsive to diverse constituencies and decentralized feedback to national policies is most likely to occur" (UNCED 1992a: 3). Moreover, the Rio Declaration on Environment and Development states that global warming is most likely to be effectively managed through the wide-scale political participation of all concerned stakeholders "at the lowest, most accessible, and policy-relevant" level (ibid.). Thus, it clearly rejects centralized, bureaucratic approaches. As noted previously, however, while this rejection would appear to lend implicit support to bioregional alternatives, nowhere does Agenda 21 explicitly call for a bioregional approach, method, or alternative.

What is less clear is precisely how effective these varied and various efforts will be. For example, while a number of "local climate protection initiatives" have begun under the aegis of Agenda 21, thus far, the highest degree of success achieved is in convening workshops comprised of local officials designed to share information and experiences, enlist local officials to develop evaluation guidelines for producing renewable energy sources, reducing energy-use, and developing novel, innovative transport and land-use programs (Parenteau 1995; FEM 1995). However, independent validation or evaluation of the achievement of sustainability has not yet been formally undertaken by any non-governmental body or organization. The following section is an attempt to outline some steps that have been taken to begin to address these very problems.

Moving toward a bioregional approach

One of the best sets of examples of an attempt to encourage a bioregional-type approach to climate-change has been occurring in the US at the behest of federal agencies. Similar activities are occurring in other countries. The Clinton Administration's 1993 *Climate Change Action Plan* promotes strengthening federal support for US state global climate-change activities. The administration has committed support for state and local initiatives, provision of revolving funds to finance the design of energy-management and retrofit programs in public buildings, assistance to monitor actions to reduce transportation conges-tion, and technical and financial support to public-service commissions to promote utility investments in energy-efficiency, demand-side management and integrated resource planning programs (Clinton and Gore 1993). One inter-esting aspect of this program is that it is designed to enhance existing state activities.

Additionally, the now-defunct Office of Technology Assessment (OTA) has recommended that federal agencies utilize state- and region-derived climatic data in order to ascertain how clouds and water vapor affect the distribution of solar energy, and thus regional variations in greenhouse gas-induced warming. OTA also endorsed a stronger state role in evaluating the societal impacts of larger atmospheric concentrations of greenhouse gases (US Congress, OTA 1993b: 5–6).

In other federal systems such as Canada and Germany, provinces and *länder*, respectively, are taking an active role in reducing greenhouse gases through setting forth regional environmental reporting and greenhouse gas reduction goals, as well as promoting comprehensive, integrated metropolitan area energy, district heating, conservation, transport and land-use planning in order to reduce greenhouse gas emissions (UNCED 1992a; FEM 1995). In Nigeria, a developing nation where the state has traditionally had little substantive authority to undertake environmental planning or regulation, several institu-tion-building activities related to climate-change are taking place. These include establishment of special state environmental committees with oversight authority (UNCED 1992a: 176–8).

In unitary systems, subsystem activities also are being encouraged. For example, France relies upon local authorities for solid and hazardous waste landfill management (related to methane control) and is developing a "partner-ship with local organizations and decentralization of responsibilities" for sustainable development and other global change-related decision-making. This partnership solicits nonbinding opinions from existing regional planning authorities and departments (UNCED 1992a: 114).

In short, states, regions and other subsystems in several countries are engaging in climate management activities. There appear to be strong motiva-tions for activity that run counter to collective action disincentives that discourage subsystem involvement in climate-change activities. These motiva-tions may or may not be strong enough to completely overcome disincentives to generate a socially optimal level of activity or to create a bioregional approach. However, the evidence points to three possible motivations that, given the right conditions, may help generate underlying political support for a bioregional-type approach to the problem of climate-change.

The first motivation is the previously noted indeterminacy of climate-change impacts. In regions where highly valued, climate-sensitive resources may be threatened by climate-change, support for subnational activities to address it is likely to be high. Such regions are likely to be less economically diversified and thus more vulnerable to climate-change impacts (Clark 1988). A second motivation is the desire to exploit climate-change activities as a tool for building support for other environmental or economic priorities. This "catalyst" motive is sometimes called "no regrets;" an action taken in pursuit of interests such as protection of jobs and the economy, but which also may have favorable, incidental impact on efforts to reduce greenhouse gases (Wells 1991; Flavin and Piltz 1989). A final motive – that points precisely to the role of underlying regional cultural factors in promoting subnational, bioregional-type activities – is regional character. Acceptance of activist and innovative environmental policies tends to correlate highly with distinct social, cultural and economic indicators. These include a tradition of political "progressivism," active environmental groups and high levels of education among the regional populace (Ringquist 1993, 1994; Lester 1994; Morandi 1992; Feldman and Wilt 1994; Jones 1991; Clark 1988; Grodzins 1983; Vig and Kraft 1994).

National policies may increase incentives and reduce disincentives, generating additional subsystem activity. In order for national policies to be appropriately designed, we must understand both the motivations that spur activities and the constraints which limit them. This understanding will better permit development of national strategies to coordinate efforts to meet national signatory commitments under international agreements.

A focus on national efforts: The case of the United States

As noted previously, an array of federal initiatives have been developed to encourage bioregional-type management of climate-change. These activities have spurred many US states to explore options for global climate-change activities. However, several states, including California, Connecticut, Missouri, Iowa and Oregon, had begun climate-change initiatives prior to the publishing of the OTA recommendations or the Clinton Administration's *Climate Change Action Plan*. Many of the activities included in these states' programs are based on energy conservation, reforestation, transportation planning, waste management and purely symbolic measures (Wilt and Feldman 1995). Because of their relative innovativeness, they are good candidates for exploring the viability of bioregionalism and climate-change.

In 1988, California became the first state to address the issue of global warming in legislation. Assembly Bill 4420 required the California Energy Commission to study potential impacts of climate-change on the state's energy supply-and-demand, economy, environment, agriculture and water resources, and make legislative recommendations to reduce greenhouse gas emissions. The California Energy Commission subsequently released *Global Climate Change: Potential Impacts and Policy Recommendations* (vols. I and II) in 1991. Assembly Bill 2360, passed in 1989, required the Office of Planning and Research to review specific provisions of the state Environmental Quality Act to determine whether programmatic measures should be changed in response to potential impacts of global change (US EPA 1991; Morandi 1992).

California has also developed other legislation and agency policies that promote reductions in greenhouse gases. The California Air Resources Board is analyzing potential demand-side measures to reduce emissions of greenhouse gases in the state's South Coast Air Quality Control District. House Bill 7325, passed in 1989, requires gas utilities to submit yearly conservation plans to the Public Utility Commission – the plans must include measurable conservation and load management conservation targets, conservation options, analyses and evaluation methods, and cost/benefit findings.

In 1989, Missouri passed a resolution creating a fourteen-member Commission on Global Warming and Ozone Depletion. The charge of the Commission was to assess Missouri's contribution to ozone depletion and form policy options to deal with the effects of those problems. In 1992, the findings of the Commission were published in the *Report of the Missouri Commission on Global Climate Change and Ozone Depletion.* Several policy actions came out of the Commission's recommendations including: a goal for a 20 percent reduction in state carbon emissions by the year 2005; a 30 percent improvement in energy conservation in public buildings; statewide minimum energy efficiency standards for new construction and renovations; soil protection planning; and development of biomass programs to provide alternative fuels.

Missouri also developed "Operation T.R.E.E. – Trees Renew Energy and Environment," as a measure to minimize global change and soil erosion while promoting energy conservation. The state Department of Natural Resources, Division of Parks and other state agencies work closely in several components of Operation T.R.E.E., including using public volunteers to plant over 50,000 trees, reforesting reclaimed coal and mineral mines, and developing educational materials to encourage more agricultural, residential and commercial reforestation. In addition, 5,000 wooded acres have been targeted for intensive management: 2,200 acres will be reforested, with the remainder slated for natural growth (i.e. rather than deliberate reforesting efforts, natural ground cover and vegetation will be allowed to flourish).

Improving a bioregional approach

Identifying regional scale impacts to governments and other institutions has been anecdotally acknowledged by some subsystem officials to be important for bioregionalism. This would enable a better understanding of the prospects for adapting regional scale sustainable economic practices and for acceptance of small-scale, decentralized, autonomous and "open" decision-making mechanisms that incorporate the preferences of numerous and diverse stakeholders. By stakeholders, we mean people who have a stake or interest, broadly defined, in regional welfare. Generally, this is taken to include government officials, private sector economic interests, and non-governmental organized groups. Ideally, however, it should also encompass average citizens and indigenous peoples ordinarily excluded from positions of power or special influence but who nevertheless suffer when resources are abused or misused, and who often lack a voice in decisions – even though they are often the closest to resources, most vulnerable to their mismanagement and sometimes best able to understand how to manage them sensibly (Feldman 1991; 1995).

NGOs can play an important role in climate-change policy development, implementation and monitoring. A promising example of a NGO that has demonstrated a capacity for sustaining underlying support of subnational climate-change activities is the International Council for Local Environmental Initiatives (ICLEI) based in Toronto. ICLEI represents local governments that are committed to sustainable development. It engages in research and demonstration activities on behalf of local and metropolitan governments worldwide and its activities are wide-ranging and include investigations of energy-use in the cities of developing countries and the relationship between population density and energy-use (Torrie 1993). Like most effective NGOs (e.g. the World Meteorological Organization, various environmental and natural resource groups), ICLEI has tended to focus on areas where it can most effectively fill a void given both contributor state interests and staff abilities. This has meant that it has paid less attention to other areas relevant to climate-change that fall outside of energy. If there is a lesson here, it would appear to be that while NGOs can play an important role in the establishment of bioregionally oriented forms of dealing with climate-change, that role may be limited to select aspects of the problem. No NGO or set of NGOs is equipped to cover all aspects of the issue. Government-sponsored programs and policies are necessary. This is an important consideration in trying to design an effective role for NGOs in sustaining bioregional climate-change efforts.

In unitary systems such as France and the UK, structural reforms that began in the early 1980s – permitting local land-use planning and discretionary use of national tax resources by local/regional governments resulted in greater empowerment of regional governments in environmental policy (Buck 1989; Feldman 1989). Unfortunately revenues also did not keep pace with expectations.

While subnational regions vary in their direct resource commitments to global change, the good news is that states – in economically developed polities – have the ability to indirectly "leverage" additional resources through imposing standards on the private sector, licensing or permitting of certain activities, or zoning (Sand 1987). How states leverage these additional resources is an important consideration in bringing about a bioregional approach to climate-change management. The Pennsylvania Energy Office (PEO), for example, funds a wide variety of energy efficiency and conservation projects, including energy technology demonstration, recycling/composting, and solar powered vehicle projects, as well as community action plans (*Global Warming and Energy Choices* 1995). California also heavily supports direct resource expenditures on energy education and technology demonstration. What California and Pennsylvania share, however, is an avid use of indirect resource leveraging through regulation and standard-setting measures that require expenditures by the private sector and, when passed through, the consumer. In California's case, state requirements for utility least-cost planning, building standards and renewable energy developments, while placing initial economic burdens on some sectors, is viewed as a long-run investment in both energy and cost savings.

Government initiated programs, however, are not necessarily compatible with a bioregional framework. While many government-initiated climate-change related programs are designed to address the complex relationships between human activities and ecosystem damage (e.g. reforestation programs, transportation and land-use planning), many resources requiring comprehensive management are themselves trans-state – or even substate – in character. Effective management of many global change-impacted resources may require conjoint programs among and between states and NGOs. River-basin management schemes of this type have been tried in the Delaware and Colorado River Basins: arrangements optimally suited for managing issues such as water supply, interbasin water-quality and energy supplies dependent on direct hydropower or cooling water for steam plants. Unfortunately, while these schemes are theoretically based on what are referred to as functional as opposed to political–territorial criteria, in practice they have often failed to exploit these criteria due to insufficient authority to overcome stakeholder self-interest and lack of perceived legitimacy to make difficult allocative choices (Feldman 1991; 1995). This problem exemplifies the misappropriation of bioregional values by some states, and the use by others of bioregional concepts of organization while retaining bureaucratic and narrow constituency-based values of management (for a discussion of this, please refer to Chapter 4 by McGinnis).

Likewise, some global change-related issues might better be dealt with by the adoption of bioregionally oriented approaches that are based on smaller management systems (e.g. watershed-based programs). Examples of global climate-change related problems amenable to this sort of management could include environmental monitoring of a number of air- and water-quality issues by citizens, as well as resource recycling programs. The aforementioned local management programs in the UK exemplify how these programs might work and, being territorially small, they might make for a more effective match between cultural and political constraints, and resource problems.

A bioregional response to climate-change lies in thinking critically about the unconventional types of institutions that might be appropriate for policy development, implementation and monitoring. Policy analysts and others should be cautious and recognize that in those cases where the motive for policy is not to promote a truly sustainable result, then the bioregional institutions that may come about will probably be more fragile than might otherwise be the case. Bioregional organizations need fiscal resources and authority to take independent action on a vast array of decisions. It is possible that, since these bioregional institutions will be competing with nonbioregions, their political legitimacy is more likely to be subject to political debate. Finally, care should be taken not to confuse conscious bioregional policies with strategies designed to attract "green" businesses, to enhance economic competitiveness, or to respond to other more palpable environmental problems. Desirable results may emerge, but their causes may be based as much on prudential societal self-interest as on loftier bioregional premises.

Conclusions

As we have demonstrated, there are numerous examples of subnational policy and program development aimed at mitigating global climate-change. These efforts have overcome the pervasive disincentives that work against such activity. However, from a bioregional perspective, these activities are still based upon the formal functional areas traditionally designated as falling within the realm of subsystems' responsibility. Further, these jurisdictions are still those that have been set up with arbitrary bureaucratic boundaries, which may or, more likely, may not have any basis in rational bioregional boundaries.

Environmental problems, such as those associated with climate-change, are perfectly suited to be addressed on a bioregional scale. However, in order for this to occur, federal and subnational governments must provide incentives for such cooperation to take place outside of traditional institutional realms. There are many examples in the US of special districts composed of areas within a state, or several states, such as regional planning commissions, federal–state interagency regional organizations (e.g. the Appalachian Regional Commission, Delaware River Basin Commission) or even federal corporations with a specific regional mandate (e.g. Tennessee Valley Authority). However, these organizations generally lack authority to formulate and implement public policies toward the environment, and furthermore, lack elected officials who are expected to be responsive to public interests (Derthick 1974; Ostrom, Schroeder and Wynne 1993; Mann 1993; Matthews 1994; McGinnis 1993). In other words, they often still require states to provide action. Moreover, while states and communities have been granted a number of incentives for developing climate-change responses under the *Climate Change Action Plan* (Clinton and Gore 1993), there has been, as yet, no explicit attempt to introduce incentives to promote development of climate-change related programs for the special district institutions cited above. As a starting point, one way that policy-makers might encourage bioregional management of climate-change problems would be to introduce such incentives as those that have been provided to states and communities under the *Climate Change Action Plan*. This would ensure that climate-change related policies in energy, natural resources management and related areas can be tailored to these nonstate jurisdictions.

States and regional political subsystems have begun to recognize their critical role in assessing the effects of climate-change, formulating mitigation plans and identifying management strategies. However, it is far from clear how climate-change can be managed on national-scale levels, much less bioregionally. While climate-change policy provides a unique example of how bioregional problems and solutions may evolve within complicated national and international decision frameworks, it remains a problematic example at best. Thus, we end close to where we began in two respects:

1 Even where climate-change responses have been undertaken at local and regional levels, identifying a decision-making focus – a set of political actors who are ultimately responsible for policy-formulation and the implementation of climate-change mandates under, say, international agreements – remains difficult. Local and regional officials respond, as we have seen, to problems that are perceived as distinctly regional or local. Where they exercise proactive authority in these areas, they do so because structural reform at the national level (e.g. UK, New Zealand) encourages them to do so, or because political and economic opportunity (e.g. US) makes it advantageous for them to do so.
2 The second problem, while most difficult to surmount, ironically gives us greater hope. The causes and consequences of climate-change are spatially and temporally ubiquitous. No single region or set of regions can, by itself, address the overall problem. Cumulatively, however, management approaches that are regionally based, integrated, and that involve affected stakeholders directly in the design of programs to address the issue may have the best chances of reducing greenhouse gas emissions, promoting adaptation policies and mitigating adverse impacts.

In essence, a bioregional approach to climate-change could be based on a network of both bioregional and even nonbioregional, but, nevertheless, subnational institutional arrangements that could tackle energy, environmental quality and natural resource issues. Table 8.1 provides a conceptualization of how a bioregional approach might be tailored to scale. Simply stated, the rationale for this matrix is the notion that many of the activities most responsible for causing global warming (e.g. energy production and consumption, deforestation, transportation choice) are influenced by actions taken at the subnational level and sometimes fall outside the traditional purview of formal regional political institutions (e.g. states, provinces).

Moreover, many of the conditions that could possibly result from global climate-change (e.g. drought, desertification, flooding, agricultural production, coastal zone and water resources management threats, threats to biodiversity and habitat) may also fall outside of the realm of these traditional subnational level institutions.

Table 8.1 Roles and responsibilities for managing global climate-change related environmental issues at various governing scales

Driving force behind climate-change/ functions to be managed	Global scale	Nation-state	Sub-state region (state, province)	Bioregion	Local entity (county/city)	Non-governmental organization
Energy-use	Transnational agreements to reduce CO_2	Setting polices and regulations for energy conservation; grants/tax incentives for conservation programs	Technical assistance; education and outreach	Managing interface between energy projects and natural systems	Building conservation and end-use efficiency programs	Demonstration projects for local energy sustainability
Deforestation	Treaties to discourage forest products in trade; debt-for-nature swaps	Preservation set-asides; enforce uniform standards	Subsidies/loans/ grants/tax incentives for replanting and decreased use of virgin timber; promoting diverse species versus mono-cultures	Coordination when management issue entails a forest region and its boundaries as well as adjacent, economically tied communities	Tree planting programs	Assistance in establishing criteria/ evaluation; research; debt-for-nature swaps
Transportation demands/needs	—	Grants; research and development (e.g. smart highways); tax incentives for car pooling; develop fuel efficiency standards	Diamond lanes; public transportation for car subsidies	Managing interface between transportation corridors and land and ecosystem boundaries	Development of car pooling programs; public transportation; bike trails	Develop fuel efficiency standards; assess government data

Land-use and recreation	Establishment of international biosphere preserves	Designation of national, historic, scenic and recreational sites	Enforcement of construction and coastal zone restriction	Assessment of threatened national parks or natural areas – stronger role needed	Education/outreach	Assess government data
Ozone-depleting compounds	International trade and regulatory agreements; ban substances	Responsible for setting policies/regulations to ensure coordination of principles and standards	Oversight; technical assistance; education	—	Education/Outreach	Assess government data
Water projects	International agency funding for projects	Develop principles and standards for environmentally appropriate development	Enforcement of regulations; technical assistance	Evaluation of watershed or river-basin characteristics; ensuring conformity of development with natural area	Enforce shoreline development practices in line with bioregional guidance	Refugee and resettlement issues; also advocate for environmentally appropriate development; impact assessment
Diminishing biodiversity	Treaties to protect ecological resources and species	Preservation/conservation programs; research and development on species breeding and reintroduction; establish set-asides	Similar to nation-state roles	Research/assessment of pressures and stresses on ecosystem diversity; creating and managing ecosystem corridors	Creating ecosystem corridors	Assess government data; track biodiversity losses/gains; research; help establish set-asides

In short, what we are proposing is a tiered approach to bioregionalism. This approach can be seen in the example of biodiversity and habitat threats. We have good reason to believe that climate-change will threaten sensitive ecological habitats by creating meteorological conditions that are not only severe, "possibly disrupting these ecosystems and the services they provide," but that have possibly irreversible effects (Watson, Zinyowera and Moss 1996). While international agreements could be developed to protect biodiversity, in part, through protocols to the framework convention on climate-change, it will require national laws and regulations to ensure that habitat destruction is forestalled.

Likewise, state and local policies encouraging wise land-use may need to be imposed. However, monitoring and management of habitat changes may best be suited for special bioregions that are organized on the basis of forest, prairie, or other landscape parameters, in conjunction with appropriate political authorities (e.g. neighboring/adjacent villages and communities) and economic patterns of livelihood (e.g. economic and noneconomic NGOs). While such imaginative arrangements pose a considerable challenge to conventional ways of organizing, they are not unprecedented. The expectation that international goals can be achieved through cooperation with local NGOs and unconventional institutions has already become part of the mode of operation for the Global Environment Facility (comprised of the World Bank, UN Development Program and UN Environment Program), especially in relation to so-called "mega-biodiversity areas" in countries such as Costa Rica (Sharma 1996). It has also become part of the lexicon, as we have seen, of local sustainable development efforts in countries such as the UK and New Zealand.

Notes

1 The distinction here is important because the earth's climate has always been subject to natural fluctuations. What is significant about global climate-change as an environmental issue today is the fear that emissions from industry, transportation, power generation, deforestation and a host of other activities – activities that increased dramatically beginning in the late nineteenth century – are trapping the sun's infrared radiation, producing a dramatic, relatively sudden increase in global average temperatures (estimated at 2–3 degrees Celsius over, perhaps, 75–100 years). Many scientists believe that this temperature increase will lead – and may already be leading to – a number of projected outcomes, including damage to coastal areas from rising sea levels, more intense tropical storms, higher incidence of heat-related disease, destruction of fragile ecosystems, depletion of freshwater, more urban smog and massive shifts in world food and forest resources production. These impacts are highly variable from region to region (Schneider 1990; MacKenzie 1990; Watson, Zinyowera and Moss 1996).

Bibliography

Anderson, T. and Leal, D. (1991) *Free Market Environmentalism*, Boulder: Westview.

Atkinson, A. and Vorratnchaiphan, C.P. (1994) "Urban Environmental Management in a Changing Development Context: The Case of Thailand," *Third World Planning Review* 16 (2): 147–70.

Berlin Communiqué (1995) *Municipal Leaders' Communiqué to the Conference of the Parties to the UN Framework Convention on Climate Change, Second Municipal Leaders' Summit on Climate Change* (March), Berlin: International Council for Local Environmental Initiatives.

Buck, S. (1989) "United Kingdom Environmental Policy," in F. Bolotin (ed.) *International Public Policy Sourcebook*, vol. 2, *Education and Environment*, Westport: Greenwood.

California Energy Commission (1991) *Global Climate Change: Potential Impacts and Policy Recommendations*, vols. I and II.

Chiang, C.H. (1994) "Malaysia Country Report," *Climate Alert: A Publication of the Climate Institute* 7 (4): 7.

Clark, W. (1988) *The Human Dimensions of Global Climate Change: A Report Prepared for the US National Research Council's Committee on Global Change*, Cambridge: Harvard University.

Clinton, W.J., President of the United States, and Gore, A., Jr., Vice President of the United States (1993) *The Climate Change Action Plan*, Washington, D.C.: The White House.

Davis, C.E. and Lester, J.P. (1987) "Decentralizing Federal Environmental Policy," *Western Political Quarterly* 40 (2): 555–65.

Derthick, M. (1974) *Between State and Nation: Regional Organizations of the United States*, Washington, D.C.: Brookings Institution.

Devall, M.S. and Parresol, B.R. (1994) "Global Climate Change and Biodiversity in Forests in the Southern United States," *World Resource Review* 6 (3): 376–94.

Deyle, R.E., Meo, M. and James, T.E. (1994) "State Policy Innovation and Climate Change: A Coastal Erosion Analogue," in D.L. Feldman (ed.) *Global Climate Change and Public Policy*, Chicago: Nelson-Hall.

Dryzek, J. (1987) *Rational Ecology: Environment and Political Economy*, Oxford: Basil Blackwell.

FEM (1995) *Environmental Policy: Local Authority Climate Protection in the Federal Republic of Germany* (March), Bonn: Federal Environment Ministry of Germany.

Feldman, D.L. (1989) "France's Environmental Policy," in F. Bolotin (ed.) *International Public Policy Sourcebook*, vol. 2, *Education and Environment*, Westport: Greenwood.

—— (1991; 1995) *Water Resources Management: In Search of an Environmental Ethic*, Baltimore: Johns Hopkins University Press.

Feldman, D.L. and Wilt, C.A. (1994) "States' Roles in Reducing Global Warming: Achieving International Goals," *World Resource Review* 6 (4): 570–84.

—— (1996) "Evaluating the Implementation of State-Level Global Climate Change Programs," *Journal of Environment and Development* 5 (1): 46–72.

Flavin, C. and Piltz, R. (1989) "A Sustainable Energy Future," *Sustainable Energy*, Washington, D.C.: Renew America.

Frost, S. (1994) "Global Ecological Change," *International Journal of Environmental Studies: Section A – Environmental Studies* 45 (3–4): 169–71.

Furuseth, O. and Cocklin, C. (1995) "Regional Perspectives on Resource Policy: Implementing Sustainable Management in New Zealand," *Journal of Environmental Planning and Management* 38: 181–200.

Global Warming and Energy Choices: A Community Action Guide (1991) Washington, D.C.: Concern.

Grodzins, M. (1983) *The American System: A New View of Government in the US*, New York: Transaction.

Grubb, M., Koch, A., Munson, F. and Sullivan, K. (1993) *The Earth Summit Agreements: A Guide and Assessment – An Analysis of the Rio '92 UN Conference on Environment and Development*, The Royal Institute of International Affairs: Energy and Environmental Programme, London: Earthscan Publications.

Haas, P. (1996) "Is 'Sustainable Development' Politically Sustainable?" *The Brown Journal of World Affairs* 3 (2): 239–46.

Jones, B.S. (1991) "State Responses to Global Climate Change," *Policy Studies Journal* 19 (2): 73–82.

Kaplan, M. and O'Brien, S. (1991) *The Governors and the New Federalism*, Boulder: Westview.

Kliendorfer, P. and Kunreuther, H. (1986) *Insuring and Managing Hazardous Risks: From Seveso to Bhopal*, New York: Springer-Verlag.

Lee, K.N. (1993) *Compass and Gyroscope: Integrating Science and Politics for the Environment*, Washington, D.C.: Island Press.

Lester, J.P. (1994) "A New Federalism? Environmental Policy in the States," in N. Vig and M. Kraft (eds.) *Environmental Policy in the 1990s*, 2nd edn., Washington, D.C.: Congressional Quarterly Press.

Linstone, H.A. (1984) *Multiple Perspectives for Decision Making: Bridging the Gap Between Analysis and Action*, New York: Elsevier.

MacKenzie, J. (1990) "Toward a Sustainable Energy Future: The Political Role of Rational Energy Pricing," *WRI Issues and Ideas*, Washington, D.C.: World Resources Institute.

Mann, D.E. (1990) "Environmental Learning in a Decentralized World," *Journal of International Affairs* 301.

—— (1993) "Political Science: The Past and Future of Water Resources Policy and Management," in M. Reuss (ed.) *Water Resources Administration in the United States: Policy, Practice and Emerging Issues*, East Lansing: Michigan State University Press.

Martin, S. (1995) "Partnerships for Local Environmental Action: Observations on the First Two Years of Rural Action for the Environment," *Journal of Environmental Planning and Management* 38: 149–66.

Matthews, O.P. (1994) "Judicial Resolution of Transboundary Water Conflicts," *Water Resources Bulletin* 30 (3): 375–83.

May, P.J. (1995) "Coerce or Cooperate? Rethinking Intergovernmental Environmental Management" (September), unpublished paper, Chicago: American Political Science Association Annual Meeting.

McGinnis, M.V. (1993) *Bioregionalism: An Ecological Approach to Administration*, Ph.D. dissertation, University of California, Santa Barbara.

McLaren, D.J. and Skinner, B.J. (1987) *Resources and World Development*, Chichester: John Wiley.

Meo, M. (1991) "Sea-level Rise and Policy Change: Land-Use Management in the Sacramento-San Joaquin and Mississippi River," *Policy Studies Journal* 19 (2): 83–92.

Missouri Commission on Global Climate Change and Ozone Depletion (1991) *Report of the Missouri Commission on Global Climate Change and Ozone Depletion* (24 September), State of Missouri.

Morandi, L. (1992) "Assessing the State Legislative Response to Global Warming," *State Legislative Report*, Denver: National Conference of State Legislatures.

Morphet, J. and Hams, T. (1994) "Responding to Rio: A Local Authority Approach," *Journal of Environmental Planning and Management* 37 (4): 479–86.

NGA (1990) *A World of Difference: Report of the Task Force on Global Climate Change*, Washington, D.C.: National Governors' Association.

Nordhaus, W.D. (1991) "To Slow or Not to Slow: The Economics of the Greenhouse Effect," *Economic Journal* 101: 920–37.

Ostrom, E., Schroeder, L. and Wynne, S. (1993) *Institutional Incentives and Sustainable Development: Infrastructure Policies in Perspective*, Boulder: Westview Press.

Parenteau, R. (1994) "Local Action Plans for Sustainable Communities," *Environment and Urbanization* 6 (2): 183–200.

Richards, L. and Biddick, I. (1994) "Sustainable Economic Development and Environmental Auditing: A Local Authority Perspective," *Journal of Environmental Planning and Management* 37 (4): 487–94.

Ringquist, E.J. (1994) "Policy Influence and Policy Responsiveness in State Pollution Control," *Policy Studies Journal* 22 (1): 25–43.

—— (1993) *Environmental Protection at the State Level: Politics and Progress in Controlling Pollution*, Amonk: M.E. Sharpe.

Sadduzzaman, M.A. (1994) "Bangladesh Country Report," *Climate Alert: A Publication of the Climate Institute* 7 (4): 3.

Sand, P. (1987) "Air Pollution in Europe: International Policy Responses," *Environment* 29: 16–20; 28.

Sari, A.P. (1994) "Indonesia Country Report," *Climate Alert: A Publication of the Climate Institute* 7 (4): 6.

Schneider, S.H. (1990) *Global Warming: Are We Entering the Greenhouse Century?* San Francisco: Sierra Club Books.

Sharma, S.H. (1996) "Building Effective International Environmental Regimes: The Case of the Global Environment Facility," *Journal of Environment and Development* 5 (1): 73–86.

Torrie, R. (1993) *Findings and Policy Implications from the Urban CO_2 Reduction Project*, Toronto: International Council for Local Environmental Initiatives.

Tversky, A. and Kahneman, D. (1974) "Judgment Under Uncertainty: Heuristics and Biases," *Science* 185: 1124–31.

UNCED (1992a) *Agenda 21 Rio Declaration Forest Principles* (drafts) New York: United Nations Conference on Environment and Development.

UNCED (1992b) *Nations of the Earth Report*, vol. 1, *National Reports Summaries*, Geneva: United Nations Conference on Environment and Development.

UNEP (1987) *Montreal Protocol on Substances that Deplete the Ozone Layer* (final act), New York: United Nations Environment Programme.

United Nations Framework Convention on Climate Change (June) (1992) New York: United Nations.

US Congress, Office of Technology Assessment (1993a) *Preparing for an Uncertain Climate: Summary*, OTA-O-563, Washington, D.C.: Government Printing Office.

—— (1993b) *Global Change Research and NASA's Earth Observing System*, OTA-BP-ISC-122, Washington, D.C.: Government Printing Office.

US Environmental Protection Agency (1991) *Project Summary: EPA/NGA Workshop on Global Climate and State Actions, December 3–4, 1990*, Washington, D.C.: Air and Energy Engineering Laboratory.

Vig, N. and Kraft, M. (1994) "Conclusion: The New Environmental Agenda," in N. Vig and M. Kraft (eds.) *Environmental Policy in the 1990s*, 2nd edn., Washington, D.C.: Congressional Quarterly Press.

Watson, R.T., Zinyowera, M.C. and Moss, R.H. (1996) *Climate Change, 1995: Impacts, Adaptation and Mitigation of Climate Change – Scientific-Technical Analyses: Contribution of Working Group II to the Second Assessment Report of the Intergovernmental Panel on Climate Change*, Cambridge and New York: Cambridge University Press.

Wells, B. (1991) *State Programs that Reduce Atmospheric Concentrations of Greenhouse Gases*, Washington, D.C.: National Governors' Association.

—— (1993) *Climate Change Mitigation: Part of a Sustainable Development Strategy* (November), Washington, D.C.: National Governors' Association.

Wexler, P. (1992) *Cool Tools: State and Local Policy Options to Confront a Changing Climate*, College Park: Center for Global Change, University of Maryland.

Williams, R.H. (1991) "Low-Cost Strategies for Coping with CO_2 Emission Limits," *Energy Journal* 11: 35.

Wilt, C.A. and Feldman, D.L. (1995) "US Commitments and Responsibilities to Reduce Global Warming: Contributions of State-Level Policies and Programs," *Environmental Challenges: The Next Twenty Years*, Washington, D.C.: NAEP 20th Annual Conference Proceedings.

Part III

Local knowledge and modern science

9 Combining science and place-based knowledge

Pragmatic and visionary approaches to bioregional understanding

Bruce Evan Goldstein

As heirs to the back-to-the-land and appropriate technology movements of the 1960s and 1970s, many bioregionalists question whether scientific experts provide the only dependable source of knowledge about natural and cultural processes (Aberley 1993; Snyder 1994; Haenke 1996). The alternative to the exclusive authority of scientific knowledge is "place-based knowledge," rooted in the complex social activities of communities located in specific places. Bioregionalists call for the movement to cultivate a "grounded, authentic, local knowledge rather than abstractions, diversity and decentralization rather than standardization and centralization" (McCloskey 1996).

This chapter argues that scientific knowledge and place-based knowledge can be combined if bioregionalists break from the prevailing view of a "realist" orientation to science – a view that holds that "Scientists are neutral transmitters of nature's wisdom, engaged in the process of building an ever-expanding corpus of facts, an ever more complicated picture of nature's processes, coming ever closer to core truths about the natural world" (Takacs 1996: 117). From this realist perspective, good science is a mirror of reality, unaffected by culture, politics, or the technical tools employed to examine nature. Alternatively, a constructivist view insists that good science is embedded in culture, but is not just a reflection of culture, since sciences set themselves apart from everyday social interaction and enter into a structured, methodologically explicit relationship with technical instruments and the elusive materiality of nature. A "constructivist" perspective on the sciences can help bioregionalists embrace what is irreplaceable in science, while sustaining a commitment to place-based knowledge.

Reliance on scientific expertise exclusively has the tendency to concentrate power in the hands of the technically and scientifically adept, transforming a democracy into a technocracy (Fisher 1990). Technocracy does not simply discount place-based knowledge but also fosters an illusion of objectivity that facilitates the transformation of moral and political questions into technical issues. However, behind the veil of objectivity, science is not empty of social meaning, but a vehicle for the importation of "dense but inadequate meanings" (Wynne 1996: 60). Human communities become just another inert property to be managed within the constraints established by powerful but "hidden" social relations. Brian Wynne illustrates this relationship in his account of how sheep farmers in Scotland understand natural variation in their environment. Wynne suggests that this knowledge differed greatly from the understanding developed by well-meaning government scientists. It was not just that the scientists were unaware of the variation in microclimate, soils and management practices between farms (although this did contribute to an inaccurate estimate of carrying capacity and the loss of market opportunities); the core difference was that the scientists asked the wrong questions, because they did not share the values that farmers placed on local adaptability and flexibility.

Unless people are encouraged to develop their own understanding of their place and have that understanding count, they are prone to becoming alienated from their bioregional community and deferring to the knowledge and authority of the technocrats. However, place-based identity is precariously situated within the patterns and flows of an increasingly global political economy. David Harvey (1990) describes the cultivation of a "sense of place" as a defensive reaction to the social impact of globalization, particularly the compression of relationships of time and space brought about through new electronic media. Bioregionalism can be understood as a response to disintegration of place-based cultural and ecological relationships that dominated the lives of pre-modern people. Hence, following Harvey, the bioregional movement described by Aberley in Chapter 2 can be defined as a movement in opposition to the values of globalization and the eradication of place-based community.

My position is that bioregionalists should not espouse the eradication of all aspects of modernity. In particular, modern science is indispensable to a contemporary bioregional movement which is situated and embedded in a global political economy. Science can address the problems associated with globalism, such as greenhouse warming, ozone depletion and loss of biodiversity. Science's shared heritage can facilitate communication between bioregionalists while place-based knowledge is particular to each culture and environment. Place-based knowledge does not always inform ecologically sound cultural practices. For example, analysis of silt cores taken from lake bottoms in Mexico suggests that erosion of agricultural soils was at least as severe before the arrival of Columbus as it was after the Spanish introduced the plow (O'Hara, Street-Perrott and Burt 1993). Bioregionalists can utilize scientific tools to maintain soil fertility as well as address a wide range of other problems and concerns. Indeed, science is so useful that the temptation is to pursue bioregionalism as an application of insights from the natural and social sciences. Such an endeavor, however, might undermine the communitarian principles of bioregional governance which were described by Aberley in his chapter, and ignore the unique insights that can be derived by nonscientific knowledge practices.

The social and cultural conditions of science and place

The publication of Thomas Kuhn's *The Structure of Scientific Revolutions* (1970) was a watershed event in the destabilizing of scientific realism. Kuhn argued that science is divided into disciplinary communities. In each scientific community, different research methods, model experiments and technical languages serve to define the questions which are significant. Communication between communities, let alone collaboration, is hampered by these methods, experiments and languages because they cannot be easily acquired: they are learned through experience rather than explicit formulation and constitute a kind of "craft knowledge." During times when scientific fields are undergoing a "paradigm shift" in response to new ideas and findings, scientists are unable to agree with their peers. Kuhn believed that the disciplines could not be sensibly integrated because science is not a seamless whole. This insight applies both between the disciplines and through time within disciplines, since the introduction of a new ideas, values and paradigms carries with it the impossibility of thinking back into what preceded it, a barrier Kuhn called "incommensurability."

Kuhn's insights should serve to restrain bioregionalists from anticipating that scientific information (from ecology, biology and biogeography, among others) can deliver a grand unified theory. Rather, bioregionalists should support a constructivist view of science. One of the clearest exponents of this constructivist view was one of the earliest, Ludwick Fleck (1935; trans. 1979), who was a practicing clinical bacteriologist. Before the Second World War, Fleck proposed that scientific disciplines were not socially autonomous, closed systems. Instead, Fleck argued that an inner circle of specialists generate new knowledge claims. These claims are evaluated by a wider community of scientists, who are members of other specialist groups in similar fields. A community of scientists – or what Fleck referred to as "thought collectives" – has an open channel to the social world that exists beyond science. Thought collectives are influenced by concepts with wide social currency. Indeed, Fleck argued that social and cultural values serve as organizing principles for the generation of new scientific ideas, which he demonstrated by showing how the Wassermann blood test for syphilis reflected the ancient folk belief that syphilitics had "impure blood." Fleck described how important *social priorities* are for the advancement of scientific knowledge. Since there is no way to purify its social setting, good science is *both* objective and cultural.

Examining the cultural content of science does not commit Fleck's constructivist position to ontological relativism or the belief that physical reality does not constrain and structure scientific observation (Hess 1997). Instead, Fleck's position can be understood as an example of "semi-realism," in which scientific theories are held to be both accounts of the real and "vehicles that encode culture-bound linguistic categories and cultural values" (Hess 1997: 61). Science does not occur in a social, cultural, or political vacuum, devoid of human values. Semi-realism makes it possible to consider science and place-based knowledge as comparable knowledge practices, rather than separated by a fundamentally different relationship to reality.

It is also important to understand that "place" is a socio-cultural construct. A sense of place is situated in time, space and context. Bioregional boundaries

definition
of sense
of place

change meaning in time and space. Each individual constructs a sense of place which is predicated on perceptions of place, and influenced by factors such as their ethnicity, social class, race, and personal and family history. There are many different ways to identify with place, some of which imply different place boundaries, and others which are held together by place-based associations and institutions, such as economic and religious affiliations. As described by Lipschutz in Chapter 6 and Thomashow in Chapter 7, instead of a single bounded place, there are as many ways to bound a place as there are individuals to define it. Hence, bioregionalists should acknowledge a constructivist view of place.

Two intellectual traditions can help us understand the changing episte-mology of bioregional places. The first draws from the 1960s work of French phenomenologist Maurice Merleau-Ponty (1962), and the second is informed by cultural/human geography and anthropology. Merleau-Ponty challenges the western philosophical assumption that human beings possess a soul, a belief that has served to justify the exploitation of nonhuman organisms (as well as other cultures, the lower classes and women) that lacked sentience and the capacity for reason. Instead of positing a sheltered self, hidden deep within the body, Merleau-Ponty envisioned a self as coextensive with the body and flesh, just as sadness is associated with a heavy feeling in the limbs and joy with a heightened sensitivity of the skin (Abram 1996). This sensuous self engages in an ongoing interchange with the entities that surround it, with each object serving as an active contributor to perceptual experience. Merleau-Ponty writes: "My gaze pairs off with colour, and my hand with hardness and softness. . . . Apart from the probing of my eye or my hand, and before my body synchronizes with it, the sensible is nothing but a vague beckoning" (Merleau-Ponty 1962: 214).

This active participation of the mind, body and earthly entities is contrary to the ideal of object–subject dichotomy supported by modern science. Objectification is based on the premise that the scientist does not participate in the sensuous world. The scientist acts in isolation, as a mere observer of reality. Merleau-Ponty recognizes the active character of sensibility. His view resonates with a bioregional commitment to place-based knowledge and participation with others. He writes:

> To return to things themselves is to return to that world which precedes knowledge, of which knowledge always speaks, and in relation to which every scientific schematization is an abstract and derivative sign-language, as is geography in relation to the countryside in which we have learnt beforehand what a forest, a prairie, or a river is.
>
> (Merleau-Ponty 1962: ix)

This concept of constituting the known world (and place) through active participation with nature and others implies that knowledge is a rooted in both individual sensation and culture. For example, a traditional farmer's perception is conditioned by the native features of the surrounding landscape, the needs of the surrounding community, the climate, topography and the soil. The farmer's understanding of the farm-as-a-place is based on the interaction of rain, sun, plants and insects, and is inextricably linked to the fertility of the soil, an understanding born of personal experience as well as through education and social learning. This social, subjective, and experiential basis of farming is not less rational than science (see Postman 1992).

Since generations living in one place should promote the acquisition of place-based knowledge, bioregionalists attempt to understand and honor the knowledge of indigenous people. The construct of indigeneity is an interesting one, and it need not be based on aboriginal inhabitation. Indigeneity is a flexible construct, and can include multigenerational Montana ranchers and New England farmers as well as Native Americans, although faith in authenticity is commonly held proportional to duration of inhabitation and lack of external influence. Indigenous knowledge is as diverse as culture itself. It includes the pronouncements of holy seers, the prayers of earth mothers, tradition, ritual and myth, the precepts of common sense, the insight acquired after years of boring and back-breaking labor, and the wisdom of the elderly.

Cultural geographers and anthropologists have shed light on how an adaptive and sustainable relationship with nature is promoted by indigenous and local knowledge. In a classic study, anthropologist Roy Rappaport (1968) showed that periodic ritual prohibitions on hunting wild pig by an indigenous community served to maintain the pig population at levels that would withstand hunting pressure. In this way, place-based knowledge is often fine-tuned for ecological and cultural sustainability. An example from agriculture is *Dreifelderwirtschaft* – the three-field system of rotating between autumn-sown grain, spring-sown grain and a fallow period of grazing that farmers in Central Europe have employed to sustain soil fertility for over 1,000 years (Crosby 1986).

These indigenous knowledge practices are not only guidelines for resource-use but also methods to sustain the social as well as the natural order. Indigeneity is often expressed in a ritualistic form that encourages compliance out of respect for tradition or fear of divine retribution. Yet it is also important to understand that indigenous knowledge-systems are more than a folk science that taps into the individual unconscious. These forms of knowledge serve to sustain the social order as well as the natural order. An early proponent of this idea was anthropologist Mary Douglas (1966), who revealed the social function of the dietary prohibitions of the Israelites that are recorded in the Old Testament book of Leviticus. Douglas argued that early Jewish culture was organized around the principle of separating the complete and the pure from the mixed and the dirty. This principle was applied to dietary restrictions by reasoning that since God had assigned the animals to their appropriate domains at creation, those that did not conform to type should not be eaten. For example, shellfish violate the archetype of water-dwelling animals since they do not have scales or swim, and snakes violate the land-dwelling animal archetype because they crawl instead of walk. Conforming with these prohibitions stabilizes society, because "the dietary laws would have been like signs which at every turn inspired meditation on the oneness, purity, and completeness of God" (Douglas 1966: 58).

Douglas cites the absence of differentiation between the world and the social individual as a hallmark of primitive thought. Echoing the ideas of anthropologist Claude Lévi-Strauss (1962), Douglas proposes that primitive peoples are unconscious of the way that their societies function, and that the world appears to them as an appendage of their social institutions. In contrast, scientific practice makes it possible to develop self-aware and secular knowledge. This dichotomy underpins a progressive account of the history of knowledge, a narrative that begins with the infancy of primitive cultures and ends at mature, modern scientific man. This story has long served to justify a colonialist program of assisting indigenous people achieve cultural maturity by replacing their old rituals and superstitions with scientific reason, modern laws and institutions (Haraway 1997a).

Differentiating place-based knowledge from scientific practice using phenomenology as well as structural anthropology disrupts the logic of this oppressive developmental hierarchy of knowledge. Place-based knowledge cannot be replaced by scientific understanding because place-based knowledge is constantly regenerated through the active participation of the individual (mind and body) with place and culture. Scientific knowledge is also the product of an active relationship between scientists, their tools, and the natural and cultural worlds. But it is not the same subject, the same place, or the same culture. The objects of scientific practice are subjected to efforts to purify them in time and space (Latour 1993).

Commitment to this form of bioregional place-based knowledge avoids the ecological determinism expressed by some bioregional writers. While Kirkpatrick Sale's *Dwellers in the Land* did more to promote bioregional thinking than any other published work, Sale suggested that culture was wholly derivative of nature. Sale wrote that a bioregion was "a place defined by its life forms, its topography and its biota, rather than by human dictates; a region governed by nature, not legislation" (Sale 1985: 43). However, there remain countless examples of cultural variation within a similar ecological context. Moreover, critics of bioregionalism suggest that Sale was not only intellectually naive in believing that nature determined culture, but that he was promoting an ideology that had served to justify racial supremacy movements, such as Hitler's Aryanism (Alexander 1990; Frenkel 1994). Sale failed to recognize the crucial role of culture in the formation of bioregional and place-based knowledge practices. By depriving people of the opportunity to define their community, an ecologically-determined bioregionalism undermines the development of the civic life that Gary Snyder (1994) asserts is a result of long-term inhabitation. Without a civic life, Snyder suggests that a culture is deprived of the opportunity to develop dynamic, experimental, creative place-based knowledge.

Science and place-based knowledge combined

Both place-based knowledge and scientific knowledge are important to bioregional theory and practice. Some bioregionalists are attentive to the need to combine both forms of knowledge. David McCloskey describes bioregionalism as drawing equally from community knowledge and the ecological sciences because: "Ecology and community are two sides of the same river of life which are being lost together. Then they must also be restored together" (1996: 1). Peter Berg and Raymond Dasmann (1978) argue that science and sensibility are part of reinhabitation. So bioregionalism should not rely on one or the other epistemological framework.

Two strategies to combine science and place-based knowledge are proposed. The first strategy is a set of pragmatic recommendations for the bioregional activist, who must compete for influence in a society that does not recognize the legitimacy of place-based knowledge. In this context, the bioregionalist should search for scientific resources that are most compatible with place-based activity. The second strategy is tailored to the functioning bioregional community. In this context, bioregionalists should support a hybridized form of knowledge that is based on both place and science. This place-based science is made possible through a recognition on the part of scientists and bioregionalists of the inescapable historical contingency of their knowledge, an idea Donna Haraway (1997b) refers to as "situated knowledge."

Bioregionalists need to embrace science in order to live in a world that is to a large degree a product of science, and is primarily understood in scientific terms (Latour 1993). Since bioregionalists live in this world, they need to communicate scientifically while thinking bioregionally. In other words, in order to initiate the complex set of changes necessary to inaugurate bioregional societies, bioregionalists need to be able to draw on the ethical core and insights of place-based knowledge in order to select the scientific knowledge they need to live and communicate by. This process is made more difficult by the persistent misconception that science is a culture-free practice that provides theories that are simply explanations of reality as well as objects (such as ecosystems or genes) that are devoid of any social or cultural content (Hacking 1983). Bioregionalists who strategically use science must learn to see beyond this misconception. A critical view of science is required, and is the first step to a constructive engagement with science that does not compromise core bioregional beliefs and practices.

There are examples of place-based groups that have successfully used modern scientific forms of knowledge. Bioregionalists can learn from the NIMBY (not-in-my-backyard) organizing experiences. Community members fighting the placement of undesirable facilities, such as toxic waste incinerators or nuclear power plants, are often accused of not understanding the relatively "small risks" they face from these technologies compared to risks they accept in their everyday lives (e.g. riding a bicycle on a busy street). Despite being belittled for lack of appreciation of scientific facts associated with public health risks, thousands of grass-roots groups have successfully opposed the siting of hazardous waste storage facilities by combining an appeal to community values with crafty use of science and their place-based knowledge.

When activists employ scientific information in political arenas and the courts, they often find that the interests and policies of the sponsors of the research are subtly embedded in existing scientific information. For example, during the Pacific Northwest spotted owl controversy of the late 1980s, conservationists could not use Forest Service databases to measure the extent of old-growth forest because the Forest Service had always classified forest type by timber stand volume and tree diameter at breast-height, instead of coarse woody debris, presence of indicator species and other dimensions of an old-growth forest. Similarly, NIMBY groups opposing the siting of hazardous facilities often find that cost/benefit analyses and toxicological studies reflect the desire of the sponsors of this research to locate the most economically efficient and ecologically harmless site rather than facilitating alternatives, such as changing production processes and consumption patterns, and instituting recycling (Fiedler 1992). With their limited technical expertise and funding, activists are often unable either to detect the bias in existing science or to redesign scientific methodologies to suit their needs.

Fortunately, NIMBY groups are not alone in their struggle to adapt existing science and generate new information. They have been assisted by the Citizens Clearinghouse for Hazardous Wastes (CCHW), a group formed in 1981 by Lois Gibbs, a community activist who was prominent in the struggle to clean up New York's Love Canal. CCHW mobilizes community groups to influence decisions about toxics that are otherwise decided in closed negotiations between experts from industry, government and occasionally national environmental groups. CCHW provides over 8,000 grass-roots groups with understandable technical information through a guidebook series, a monthly and quarterly magazine, and (through the CCHW library) files and databases. CCHW policy is not to run the campaign or provide testimony for their member groups, but to enable them to do it themselves by helping them develop action strategies, research their opponents and develop technically rigorous arguments.

A central part of CCHW's strategy is to facilitate the development of new sources of information. They help organizations conduct community health surveys. They also organize their members to lobby for information resources that no single community could acquire through its own effort. In addition, CCHW has helped develop and distribute studies documenting a correlation between hazardous waste siting and the location of people of color, low-income, rural, Catholic, elderly and people without college education (CCHW 1997).

Now let's consider the case of a functioning bioregional community. Here, the public role of place-based knowledge need no longer be confined to serving as a moral guide for selecting and shaping scientific inquiry. The challenge is to bring place-based knowledge into the public realm while not sacrificing the manifold advantages of science – in other words, to create a place-based science. This is a two-step process. First, there should be a transformation in the knowledge-making practices of scientists and bioregionalists. Place-based science should be developed within new scientific institutions that can facilitate a productive dialogue between scientists and their place-based counterparts. A scientist needs to remain a dependable witness to the creation of new scientific knowledge while being attentive to the coevolution of science and society, as well as the particular interaction between knowledge making and place-making. This reflexive procedure is a difficult task because of the requirements of viewing scientific information as simultaneously real and historically contingent. As Haraway says:

> In biology, one would be a fool to say "I don't like the idea of the organism because it seems that there are these inherent relationships of domination built in. It must be just made up." Then you are carted off to the loony bin. So how do you simultaneously take the knowledge seriously while knowing that knowledge doesn't have to be that way . . . you are inside of it while at the same time knowing that the world really could be otherwise. The historicity is irreducible, and there is no way that the positive knowledge sediments out as a kind of insoluble precipitate in the bottom of the test tube, never to be dissolved again.
>
> (Haraway 1997a)

Scientists should embrace the idea that they produce work that is, in Haraway's words, "reliable, partly shareable, trope-laced, worldly, accountable, noninnocent knowledge" (Haraway 1997b: 138). This recognition has to become rooted in their scientific practice, not parceled off as a topic for discussion among ethicists. A situated perspective is difficult to embrace because of the widespread belief among scientists that science is a unitary, universal and timeless practice. Surrendering this notion will make them more capable of recognizing the potential legitimacy of place-based knowledge claims.

Engaging in place-based science requires that bioregionalists interact with community members in unaccustomed ways. To accomplish this requires face-to-face encounters between scientists and community members where they can exchange opinions and information, and develop a common knowledge base. This public forum should be capable of not only considering data but also negotiating new research methods and procedures.

One set of guidelines to accomplish this has been developed by Judith Innes (1995) using the philosophy of Jurgen Habermas (1984). Innes describes how agreement between groups and individuals who hold diverse values and beliefs can be facilitated through dialogue and negotiation in small interacting groups. Through discussion, participants sustain a critical perspective toward knowledge claims, seeking to uncover what powerful interests they may conceal. This process can help participants become aware of the assumptions and political interests embedded within knowledge; allowing them to develop a better understanding of nature and culture while increasing trust and communication that makes coordinated action possible. One of its chief advantages is that it doesn't fossilize place-based knowledge by isolating it from the sensitive, responsive community that gave it meaning and vitality. This approach does not locate place-based knowledge beyond criticism but allows it to be open to change through negotiation.

Case histories that illustrate the development of place-based science are uncommon. Scientists and native people developed a shared knowledge-base using communicative planning principles along the Mackenzie River Valley in the Arctic region of Canada's Northwest Territories, the territory of the Gwich'in people (Raygorodetsky 1996). The Gwich'in Renewable Resources Board and the Gwich'in Social and Cultural Institute set out to systematically collect and organize local knowledge, motivated by a concern that the scientific foundation for sustainable resource management is weak and indigenous traditions are disappearing because they are not getting passed on through traditional oral and on-the-land teaching. Researchers interviewed traditional people and accompanied them for two to four weeks during their hunting trips, gathering information which was taken back to a group of agency staff and knowledgeable elders, who reviewed the manuscripts for accuracy. The information gathered from the project was distributed throughout Gwich'in communities, and the project organizers hope to combine this information with scientific measures within a community-based environmental monitoring system.

The Gwich'in people developed place-based science to strengthen their claims to their traditional lands, disseminate traditional practices among their people, and ensure that their place-based knowledge is embedded inside the agencies that increasingly are responsible for managing their landscape. Applications of this kind of hybrid knowledge are more difficult to facilitate among communities who have not developed ties to their land over generations and whose cultures are deeply inflected by science. However, there are a few promising attempts. The Willapa Alliance, a community-based conservation organization representing the many diverse interests among the inhabitants of Willapa Bay, Oregon, has designed a series of innovative assessments of environmental sustainability (Ecotrust 1995; Backus 1996). Their efforts in Willapa Bay are bioregional in scope, intended to preserve the ecological health of a 600,000 acre watershed as well as sustain the livelihoods of 20,000 residents of the area who largely depend on logging, oystering, farming, fishing, tourism and other ecologically dependent activities. Drawing on the technical savvy of their sponsoring nonprofit group Ecotrust and its technical affiliate Interrain Pacific, The Alliance attempted to incorporate place-based knowledge whenever possible in the construction of their geographic informations system (GIS) database, for instance by bringing satellite imagery to oyster gatherers in order to identify the location of the most productive oysterbeds (Backus 1996). Once completed, this database was reproduced on CD-ROM and distributed to community members and institutions. In addition to this community mapping effort, the Alliance has organized a Salmon Walk and stream adoption program involving educators, students and other local citizens in natural resource issues, and an education program called "The Nature of Home" that links educators and students to their ecosystem, with the specific intention of reinforcing their sense of place.

These successes underscore the necessity of having bioregional institutions in place in order to be able to facilitate the creation of place-based science. Existing democratic institutions such as legislatures and other arenas of pluralist struggle are not well suited to facilitating consensus agreements. These pluralist institutions are rarely even capable of promoting agreement on common terms of reference between political adversaries. Instead, these forms of governance allow contending interests an opportunity to influence public policy, using whatever form of rationality they find convincing or useful. Government agencies do not provide viable bioregional forums for developing knowledge resources, because they abide by a model of scientific rationality that ignores place-based knowledge (Williams and Matheny 1995). Finally, private market-oriented institutions conform to dominant interests that are antagonistic toward bioregional objectives. Without an interest in altering fundamental inequities in the distribution of power and resources, mediation efforts are reduced to pretentious and largely ineffectual symbolic exercises that may actually diminish the capacity for bioregional governance (Innes 1995).

Conclusion

While place-based knowledge has become recognized and respected, technological change has facilitated the dissolution of regional cultures and consolidated the power of global resource and capital markets (Castells 1983). These global changes undermine a communitarian commitment to decentralized and participatory politics, soft technology and community media, and markets oriented toward meeting local human needs. Under this onslaught, bioregional efforts that rely on place-based knowledge may illustrate Lewis Mumford's rueful observation that regionalism is reactionary, an expression of our desire to find "refuge . . . against turbulent invasions of the outside world" (Mumford 1934: 292).

Rescuing bioregionalism from becoming an expression of rootless despair requires that bioregionalists develop knowledge practices that utilize place-based knowledge and embrace scientific reason as sources of insight. The product of this union, place-based science, can provide the best description of reality and still be rooted in history and culture. As bioregionalist Pat Mazza wrote:

> Intuition and instinct form as much a basis for bioregional epistemology as number and experiment. Use of all these tools, dismissing none, is the integrative revolution that is bioregionalism. If the result is shifting and processual, so is the observable universe.
>
> (Mazza 1997)

168   *Bruce Evan Goldstein*

Acknowledgments

Special thanks to Mike McGinnis, a tenacious editor and insightful scholar, who employed both science and place-based knowledge to greatly improve this chapter. I also thank Dwight Barry, Janis Birkeland, Fred Cagle, Donna Haraway, Alie Jones, Pat Mazza, Vicki Nichols, Ashwani Vasishth, Bryan Norton and two anonymous reviewers for comments on earlier drafts of this chapter.

Bibliography

Aberley, D. (1993) "How to Map Your Bioregion: A Primer for Community Activists," in D. Aberley (ed.) *Boundaries of Home: Mapping for Local Empowerment*, Gabriola Island: New Society Publishers.

Abram, D. (1996) *The Spell of the Sensuous*, New York: Pantheon.

Alexander, D. (1990) "Bioregionalism: Science or Sensibility," *Environmental Ethics* 12: 161–73.

Backus, E. (1996) personal communication at the offices of Ecotrust, Portland, Oregon.

BANA (1997) Bioregional Association of the Northern Americas: membership application, c/o Planet Drum Foundation, PO Box 31251, San Francisco, CA 94131.

Beck, U. (1994) "The Reinvention of Politics: Toward a Theory of Reflexive Modernization," in U. Beck *et al.* (eds.) *Reflexive Modernization: Politics, Tradition and Aesthetics in the Modern Social Order*, Stanford: Stanford University Press.

Berg P. and Dasmann, R. (1978) "Reinhabiting California," in P. Berg (ed.) *Reinhabiting a Separate Country: A Bioregional Anthology of Northern California*, San Francisco: Planet Drum Foundation.

Castells, M. (1983) "Crisis, Planning and the Quality of Life: Managing the New Historical Relationships Between Space and Society," *Environment and Planning D: Society and Space* 1 (1): 3–22.

CCHW (1997) *Citizens Clearinghouse for Hazardous Wastes: Center for Health, Environment and Justice Home Page* <http://www.essential.org/cchw/>

Crosby, A.W. (1986) *Ecological Imperialism: The Biological Expansion of Europe, 900–1900*, Cambridge: Cambridge University Press.

Davidoff, P. (1965) "Advocacy and Pluralism in Planning," *Journal of the American Institute of Planners* 31 (4): 331–38.

Douglas, M. (1966) *Purity and Danger: An Analysis of the Concepts of Pollution and Taboo*, New York: Routledge.

Drysek, J. (1990) *Discursive Democracy*, Cambridge: Cambridge University Press.

Ecotrust (1995) "New Bearings: Conservation-based Development in the Rain Forests of Home," Portland: Ecotrust.

Feenberg, A. (1995) "Subversive Rationalization: Technology, Power and Democracy," in A. Feenberg and A. Hannay (eds.) *Technology and the Politics of Knowledge*, Bloomington: Indiana University Press.

Fiedler, J. (1992) "Autonomous Technology, Democracy and the NIMBYs," in L. Winner (ed.) *Democracy in a Technological Society*, Boston: Kluwer.

Fisher, F. (1990) *Technocracy and the Politics of Expertise*, Newbury Park: Sage Publications.

Fleck, L. (1979) *Genesis and Development of a Scientific Fact*, Chicago: University of Chicago Press.

Forester, J. (1993) *Critical Theory, Public Policy and Planning Practice: Toward a Critical Pragmatism*, Albany: State University of New York Press.

Frenkel, S. (1994) "Old Theories in New Places? Environmental Determinism and Bioregionalism," *Professional Geographer* 46 (3): 289–95.

Habermas, J. (1984) *The Theory of Communicative Action*, vol. 2, *Lifeworld and System: A Critique of Functionalist Reason*, Boston: Beacon Press.

Hacking, I. (1983) *Representing and Intervening: Introductory Topics in the Philosophy of Natural Science*, Cambridge: Cambridge University Press.

Haenke, D. (1996) "How to Be a Bioregionalist (overview)," San Francisco: Distributed by the Research and Electronic Committee of BANA.

Haraway, D.J. (1997a) comments made during Winter Quarter Science Studies Seminar, History of Consciousness Department, University of California, Santa Cruz.

—— (1997b) *Modest_Witness@Second_Millennium.FemaleMan©_Meets_OncoMouse™: Feminism and Technoscience*, New York: Routledge.

Harding, S. (1992) "After the Neutrality Ideal: Science, Politics and Strong Objectivity," *Social Research* 59 (3): 567–87.

Harris, T. *et al.* (1995) "Pursuing Social Goals through Participatory Geographic Information Systems: Redressing South Africa's Historical Political Ecology," in J. Pickles (ed.) *Ground Truth: The Social Geography of GIS*, New York: Pergamon.

Harvey, D. (1990) "Between Space and Time: Reflections on the Geographical Imagination," *Annals of the Association of American Geographers* 80: 418–34.

—— (1996) *Justice, Nature and the Geography of Difference*, Oxford: Blackwell.

Hayles, N.K. (1995) "Searching for Common Ground," in M.E. Soule and G. Lease (ed.) *Reinventing Nature?* Washington, D.C.: Island Press.

Hess, D. (1997) *Science Studies*, New York: New York University.

Innes, J.E. (1995) "Planning Theory's Emerging Paradigm: Communicative Action and Interactive Practice," *Journal of Planning Education and Research* 14 (3): 183–9.

Interrain Pacific (1996) "Understanding Patterns of Change," Portland: Interrain Pacific.

Kuhn, T.S. (1970) *The Structure of Scientific Revolutions*, 2nd edn., Chicago: University of Chicago Press.

Latour, B. (1993) *We Have Never Been Modern*, Cambridge: Harvard University Press.

Lévi-Strauss, C. (1962) *Totemism*, Boston: Beacon Press.

Lindblom, C.E. (1959) "The Science of Muddling Through," *Public Administration Review* 19 (2): 79–99.

Lowie, I. (1990) *The Polish School of the Philosophy of Medicine*, Boston: Kluwer Academic Publishers.

Lukacs, S. (1971) *History and Class Consciousness*, London: Merlin Press.

Marx, K. (1978) "Economic and Philosophic Manuscripts of 1844," in R.C. Tucker (ed.) *The Marx–Engels Reader*, 2nd edn., New York: W.W. Norton.

Mazza, P. (1997) personal communication.

McCloskey, D. (1996) "Ecology and Community: The Bioregional Vision," *Cascadia Planet* <http://www.tnews.com/text/mccloskey2.html>

Merleau-Ponty, M. (1962) *Phenomenology of Perception*, New York: Routledge and Kegan Paul.

Miller, M.S. (ed.) (1993) *State of the Peoples: A Global Human Rights Report on Societies in Danger*, Boston: Beacon Press.

Morrison, P.H. *et al.* (1991) *Ancient Forests in the Pacific Northwest: Analysis and Maps of Twelve National Forests,* Washington, D.C.: The Wilderness Society.

Mumford, L. (1934) *Technics and Civilization,* New York: Harcourt Brace.

O'Hara, S.L., Street-Perrott, F.A. and Burt, T.P. (1993) "Accelerated Soil Erosion Around a Mexican Highland Lake Caused by Pre-Hispanic Agriculture," *Nature* 362: 48–51.

Ostrom, E. (1990) *Governing the Commons: The Evolution of Institutions for Collective Action,* Cambridge: Cambridge University Press.

Ozawa, C.P. (1991) "Consensus Based Approaches to Handling Science," in C. Ozawa (ed.) *Recasting Science: Consensual Procedures in Public Policy Making,* Boulder: Westview.

Porter, T.M. (1995) *Trust in Numbers: The Pursuit of Objectivity in Science and Public Life,* Princeton: Princeton University Press.

Postman, N. (1992) *Technopoly: The Surrender of Culture to Technology,* New York: Knopf.

Rappaport, R.A. (1968) *Pigs for the Ancestors,* New Haven: Yale University Press.

Raygorodetsky, G. (1996) "Gwich'in Environmental Knowledge Project," *Cultural Survival Quarterly* 20 (1): 15.

Sale, K. (1985) *Dwellers in the Land: The Bioregional Vision,* San Francisco: Sierra Club Books.

Snyder, G. (1994) "Coming In to the Watershed," in D. Aberley (ed.) *Futures by Design: The Practice of Ecological Planning,* Gabriola Island: New Society Publishers.

Sokal, A.D. (1996) "Transgressing the Boundaries: Toward a Transformative Hermeneutics of Quantum Gravity," *Social Text* 46–7: 217–52.

Susskind, L., Bacow, L. and Wheeler, M. (1983) *Resolving Environmental Regulatory Disputes,* Cambridge: Schenkman.

Takacs, D. (1996) *The Idea of Biodiversity: Philosophies of Paradise,* Baltimore: Johns Hopkins Press.

Western, D. (1984) "Conservation-based Rural Development," in F.R. Thibodeau and H.H. Field (eds.) *Sustaining Tomorrow: A Strategy for World Conservation and Development,* Hanover: University Press of New England.

Whitehead, A.N. (1925) *Science and the Modern World,* New York: Macmillan.

Williams, B.A. and Matheny, A.R. (1995) *Democracy, Dialogue and Environmental Disputes: The Contested Language of Social Regulation,* New Haven: Yale University Press.

Williams, J. (1996) personal communication at the Nature Conservancy, Steamboat Springs, Colorado.

Winner, L. (1995) "Citizen Virtues in a Technological Order," in A. Feenberg and A. Hannay (eds.) *Technology and the Politics of Knowledge,* Bloomington: Indiana University Press.

Wynne, B. (1989) "Establishing the Rules of Laws: Constructing Expert Authority," in R. Smith and B. Wynne (eds.) *Expert Evidence: Interpreting Science in the Law,* London: Routledge.

—— (1996) "May the Sheep Safely Graze? A Reflexive View of the Expert-lay Knowledge Divide," in S. Lash *et al.* (eds.) *Risk, Environmental and Modernity: Towards a New Ecology,* London: Sage Publications.

10 Addressing the conservation conundrum in Mesoamerica

A bioregional case study

Thomas T. Ankersen

Few global phenomena rival the final closure of the Isthmus of Panama for bioregional high drama. The earth's tectonic theatrics spawned a bioregion whose global importance far outweighs its small size. The closure of the isthmus created an oceanic dam that redirected the warm waters of the Gulf Stream north, making Northern Europe habitable, while launching a new era of marine speciation in two oceans (Ross 1996). Conversely, the land bridge between North and South America, often referred to in its bioregional context as the Mesoamerican Biological Corridor, created a migratory corridor for flora and fauna – and a route of human passage for the colonists (Stehli and Webb 1985). The legacy of colonialism has threatened many of the original inhabitants of all taxa with extinction and resulted in the political and biological fragmentation of the Mesoamerican Bioregion.

More recently, conservation biologists and policy-makers have launched an unprecedented effort to "put Humpty Dumpty back together again." Regional-scale policy instruments, binding international agreements among the seven tiny nations of Central America, expressly recognize the significance of the Mesoamerican Biological Corridor and obligate the signatories to make appropriate efforts to protect and restore this area (Ankersen 1994). The corridor comprises a grouping of existing and proposed protected areas and linkages between areas that spans the seven nations of Central America and extends into North America (through Mexico) and South America (through Colombia). Figure 10.1 provides a conceptual illustration of the proposed corridor.

Biologists refer to this region as Mesoamerica, in order to blur the political distinction between the continents. In addition to its biological significance, the corridor encompasses a wide variety of traditional and not-so-traditional indigenous homelands. The agreements represent one of the most explicitly stated binding bioregional commitments in international environmental law. At the national, subnational and community level a remarkable range of bioregionally based experiments have been initiated to protect and connect discrete components of the corridor.

This chapter describes the "conservation conundrum" in Central America. The conservation conundrum is the tension between the political and economic needs of modern society and the values of conservation, and protection of

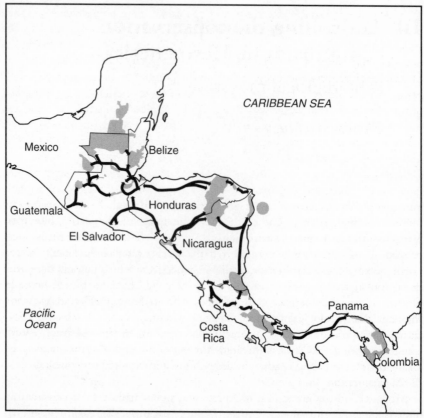

Figure 10.1 Conceptual representation of potential landscape linkages in Central America

indigenous self-determination and biodiversity. Since bioregionalism combines the values of cultural and ecological diversity, bioregional thinking may be one mechanism to address and reconcile this tension. However, bioregionalism does not appear frequently in the literature of international conservation, and it is unheard of in the conservation literature of Mesoamerica. Nonetheless, bioregionalism can serve as a useful perceptual lens from which to view a variety of conservation experiments under way in Mesoamerica. In Mesoamerica, practitioners and theorists from the international conservation movement have been advancing bioregional precepts beyond their developed-nation counterparts. This essay describes some of these efforts.

The conservation conundrum in Mesoamerica

The conservation conundrum describes the interrelated global problematic of reconciling conservation, development and local (indigenous) self-determination. It reflects the difficulties facing policy-makers and conservationists, who are seeking to ensure the maintenance of the earth's natural heritage in a world of competing values and diminishing places. Simply put, the capacity to set aside sufficiently large areas of land for the purpose of preserving biological diversity has proved to be illusive. The conservation conundrum in Mesoamerica reflects the many problems of preserving biodiversity in many places. It may, however, be more pronounced in Mesoamerica due to the nature of the ecosystems that are in the bioregion, the number and size of the political sovereignties that cut across the bioregion, and the complex social and cultural issues related to poverty, indigenous aspirations and the political and social unrest in the region.

In Mesoamerica, as in other parts of the world, forested ecosystems are inhabited. Indeed, most neotropical forests in this region have coevolved with humans since humans arrived on the continent over 10,000 years ago (Stocks 1995). Herlihy (1996) estimates that as much of 85 percent of the total protected area in Central America is occupied or used by indigenous and local peoples. Most of this is represented by the large border areas that comprise the cornerstones of the Mesoamerican Biological Corridor. In many cases, the present inhabitants represent vestiges of the original indigenous inhabitants. In other cases, the inhabitants represent the colonial aspirations of a landless poor. In either case, reconciling the needs of these inhabitants and the self-sustainable features of the landscape is a major preoccupation for the international conservation community. In Central America, there are now more than four hundred legally declared protected areas, and nearly as many proposed new ones.

While Western conservationists recognize indigenous groups and local people as allies, they do not always trust their stewardship capacity, particularly under modern circumstances (Redford 1991). Indigenous land stewardship has been challenged by a variety of complex, interrelated and contemporary factors that include: diminishing open space in which to range, improved technology to extend range and increase hunting and gathering efficiency, higher fertility rates and longevity, entry into the cash economy, and diminishing transfers of indigenous knowledge. Conversely, indigenous groups and their advocates remain suspicious of programs and policies that prioritize the needs of "nature parks" over the needs of the landscape's inhabitants. These policies and programs employ indigenous aspirations to declare protected areas while restricting indigenous development opportunities on their homelands (Herlihy 1990; Stocks 1995; Nietschmann 1995).

Mesoamerican conservation reflects this growing tension between *parquistas* (advocates nature parks) and the needs of local peoples (Stocks 1995). This can be exemplified in the reaction of a coalition of indigenous and local peoples to an initial proposal to the World Bank's Global Environmental Facility, which would designate and map the Mesoamerican Biological Corridor. In a strongly worded letter to the Central American Commission on Environment and Development, broadcast on the internet, the coordinator for Central America of the Indigenous Peasant and Afro-American Coordination for Community Agroforestry opposed the corridor project, accusing the Central American

Commission of failing to involve indigenous groups, and proposing land acquisition and displacement of Indians and peasants. In response, policy-makers sought to assuage these concerns by suggesting a name change – from the Mesoamerican Biological Corridor to the Mesoamerican Ecological Corridor – a symbolic gesture intended to indicate that humans remain an integral part of the corridor. Indeed, for most of the corridor that is a foregone conclusion. Indigenous groups and local peoples are demanding rights of self-determination and political autonomy coupled with the economic development aspirations that challenge the traditional role of the nation-state. This is particularly true in Mesoamerica where top-heavy, bureaucratic and state dominated models of governance are often enforced by oppressive military regimes. Even as the nations in the region experiment with democratic processes, political devolution to more ecologically appropriate scales of governance remain imperfect "works in progress." Communitarian demands for devolution and self-determination challenge the protected areas that have given comfort to conservationists accustomed to the role of national bureaucracies in reserve management. National governments have responded to the conservation conundrum by granting a degree of autonomy and control over natural resources within nature parks to indigenous and local people.

Yet conservationists and indigenous groups are facing pressure from recent moves in economic globalization as corporate interests seek to exploit surface and subsurface natural resources from areas that have been "earmarked" for protection. The frequent overlap of oil, minerals, timber, fisheries concessions, water resource development and transportation infrastructure projects intensify the tensions endemic to the region's conservation conundrum. Nature parks that include indigenous inhabitants are important areas that include a number of natural resources which are of interest to governments and local peoples anxious for economic development, and who have large debts to foreign countries and the World Bank. During the last two decades, the exploitation of natural resources in these areas has been suppressed by the civil strife that has characterized the Central American region (Neitschmann 1990). However, with the recent peace accord in Guatemala, the region's last declared civil war has ended. The irony is that peace in the region will undoubtedly bring about a resource development boom, which will exacerbate the region's conservation conundrum.

The overlap of protected areas, indigenous homelands and resource concessions

There exist a growing number of conflicts over land-use and the control of resources within the protected areas and indigenous homelands that make up the Mesoamerican Biological Corridor. An era of park building has come to an end in Mesoamerica. More recently, a political reaction to the parks movements has set in, and many of the established parks are suffering from attrition. Indigenous groups are suing governments to establish their land-rights, to gain control of resources and development opportunities, and sometimes to disestablish protected areas, which some view as trappings of the colonial past (Stocks 1995; Herlihy 1996). Perhaps in deference to indigenous development and self-determination aspirations, the International Union for the Conservation of Nature (IUCN) dropped the term "anthropological reserves" from its list of protected-areas categories (IUCN 1994). Refugees from the region's civil wars are invading lands located in protected areas and indigenous homelands. Multinational mining, timber and petroleum interests are making overtures to debt-ridden governments for precious resources that exist within park boundaries. Transboundary resource conflicts are confounding efforts at cooperation across borders (Meyers 1996). Moreover, the region's network of border protected areas, the crucial bioregional building blocks of the corridor have become *de facto* routes for illegal trafficking in timber, wildlife and drugs.

To be truly bioregional, each component of the Mesoamerican corridor must confront the biophysical, economic and political realities of the conservation conundrum. There are several key bioregional components of the Mesoamerican Biological Corridor: the Maya Forest shared by the nation-states of Mexico, Guatemala and Belize, and the homeland to the several Maya groups and a nonindigenous "forest society" of activists; the bi-national Mosquitia Corridor shared by the nation-states of Honduras and Nicaragua and homeland for the Moskito, Sumu, Garifuna and Pesch; and the Darien Gap shared by Panama and Colombia and homeland to the Emberra, Wounan and Kuna indigenous societies.

Sacred places: The case of the Maya Forest

The Maya Forest is a protected area in Mexico, Guatemala and Belize, and represents the largest contiguous block of tropical forest north of the Amazon. The Maya Forest extends the Mesoamerican Biological Corridor into North America. It is home to the forest-dwelling Maya in troubled Chiapas, Mexico, known as the Lacandones. The Lacandones practice a resilient form of agriculture that is believed to be key to the cultural and ecological sustainability of the Maya Forest (Nations 1988). The Lacandones have an exclusive right to inhabit a large protected area known as the Montes Azules Biosphere Reserve. This reserve lies within the political boundaries of the *Lacandon Communidad*, which is governed by the customary law of the Lacandones. Their cultural practices could offer great hope for the long-term sustainability of the Lacandon Forest. However, the reserve is also a refuge for Zapatista rebels, and it is being invaded by larger Tzeltal and Chole Maya, who do not maintain the forest stewardship practices of the Lancandones. Current estimates are that the Montes Azules Biosphere Reserve has lost a third of its forest (Guillen-Trujillo 1995).

On the other side of the border, the expansive Maya Biosphere Reserve represents Guatemala's ambitious effort to reconcile the needs of humans and the landscape. In the reserve's multiple-use zone, a nonindigenous "forest society" harvest chicle (a gum base used in chewing gum), xate (an ornamental palm leaf) and other nontimber forest products from the Maya Forest (Schwartz 1990). This forest society faces an uncertain future in the face of colonization, refugee resettlement, pressure for commercial logging, and oil concessions. A recent analysis of the biosphere reserve based on satellite imagery shows a deforestation rate of 0.2 to 0.3 percent per year, while its "buffer zone" is disappearing at a rate of between 5 and 6 percent per year (Sader 1996). In Laguna del Tigre National Park, one of the reserve's *core* zones and Central America's largest wetland, a World Bank-financed oil road into the reserve provides a conduit for colonization and development. When the government sought to remove some colonists from Laguna del Tigre, officials were taken hostage by colonists claiming the land was owed under the terms of a recent peace accord.

Belize enjoys a greater modern cultural affinity with the Caribbean and shares the natural and cultural heritage of the Maya Forest. Not unlike the other components of the corridor, this area suffers from a number of conservation conundrums. The Mopan and Kekchi Maya in Southern Belize's Toledo District are asserting land-claims over protected areas based on English common-law theories of aboriginal title, which have been successfully asserted by their counterparts in Australia and Canada (Berkey 1994). At the same time, there have been recent allegations of secret deals between the government of Belize and Asian logging interests on the Maya homelands, acquiesced in by a local conservation organization. These concessions are on national forest reserves that overlie the communal lands that are the subject of the unresolved Maya land-claims in Belize. The Toledo Maya contend they practice sustainable forms of subsistence agriculture, hunting and fishing, and believe the forest lands are *sacred places*. In addition, the Interamerican Development Bank is studying road improvements in the region – improvements that an environmental and social impact statement concludes would have negative effects on the Toledo Maya communities and their habitat. Among other things, improved transportation could exacerbate the flow of illegal immigrants from Guatemala (Stone 1995).

Preserving homelands: The case of the Mosquitia Corridor

The Mosquitia Corridor is a remote and protected region that straddles the border between Honduras and Nicaragua (Herlihy 1996). It is an essential component of the Mesoamerican Biological Corridor with an uncertain future. The region includes three large protected areas – the Bosawas Natural Resource Reserve in Nicaragua, the Rio Platano and Tawahka Asangni Biosphere Reserves in Honduras. These reserves are the homelands for several indigenous societies. The Rio Platano Biosphere Reserve covers 5,000 square kilometers of the Mosquitia Corridor, a term for the broad belt of forest and indigenous homeland between Honduras and Nicaragua. The reserve is the homeland for the Miskito, Garifuna and Pech indigenous groups. The land is owned by the state and managed by conservation groups. The smaller

Tawahka Asangni Biosphere Reserve lies near the center of the Mosquitia Corridor, and is the homeland for the Tawahka Sumu, a small indigenous group that utilizes virtually all of the reserve for subsistence hunting, fishing and some agriculture. The Tawahka have proposed a novel arrangement for indigenous management of the reserve based on UNESCO biosphere reserve principles (Wilbur 1996).

The Bosawas Natural Resource Reserve evolved from the Nicaraguan Civil War. In an effort to deter Moskito resistance to its programs, the Sandanista government divided the Atlantic Coast region of Nicaragua, largely Moskito, into two large "autonomous regions" for governance purposes. However, the degree of autonomy from the central government that these regional governments possess remains uncertain. The Chamorro government set aside three large reserves in the Atlantic Coast, including the Bosawas Natural Resource Reserve in the Mosquitia Corridor. It is the largest protected area in Nicaragua and occupies 6 percent of the country's surface area. The creation of Bosawas by the government was considered an infringement on the region's constitutionally guaranteed territorial autonomy. The reserve also overlaps Moskito and Sumu (Mayanga) homelands. The area is rich in natural resources, including timber and gold, that are viewed as a means to help Nicaragua escape the grinding poverty left by the war. In addition, Nicaragua is seeking land for repatriation of ex-soldiers and refugees from the war, and several new communities have been established on the fringes of the Bosawas forest.

With the assistance of international organizations, indigenous peoples in the Mosquitia corridor are seeking to legalize their communal homelands, including lands within the region's protected areas. This has been complicated by conflicting claims between indigenous groups and subgroups, and by colonist invasions. Moreover, development interests are seeking resource concessions, sometimes with the support and acquiescence of factions within groups who have entered the region's cash economy. In 1995, the Mayanga Indian Community of Awas Tingi submitted a claim to the Inter-American Commission on Human Rights against the Government of Nicaragua for failing to secure the Awas Tingi's land rights (Anaya and Castellón 1995). The petition contends that the Nicaraguan government was withholding title while preparing to award a long-term logging concession for timber harvesting on Mayanga homelands. The government claims that because of conflicting claims, a "comprehensive solution" to indigenous land-tenure in the region must be sought. Previously, the Awas Tingi had successfully negotiated a tripartite agreement with the Nicaraguan government and a Nicaraguan concessionaire to log communal lands (Anaya and Crider 1996). At the heart of these claims is the political and economic control of the area's natural resources; and they reveal the government's unwillingness to sacrifice the future use of natural resources by granting control of these areas to indigenous communities (Grimes 1993).

The Darien Gap

The Darien Gap, which is an area between Panama and Colombia, may be the single most important bioregional component in the Mesoamerican Biological Corridor. The Gap draws its name because it is the missing link in the Pan-American Highway, a human migration corridor that could link North and South America with ground transportation. This area's dense forests also serve as a "filter" for intercontinental speciation (Stehli and Webb 1985). Only 107 kilometers remain to fulfill the dreams of many highway engineers to have a continuous road network from Alaska to Tierra del Fuego (Korten 1994). The road was nearly completed by 1963, but an outbreak of Hoof and Mouth Disease in South American cattle convinced the United States to halt construction, explicit recognition of filter effect. Presently, the gap remains a dense tropical forest inhabited by the Emberra and Wounan indigenous peoples, who enjoy autonomy on their homeland.

Overlapping the indigenous homeland and the proposed route for the highway is Darien National Park, an internationally designated World Heritage Site. Conservationists and the Emberra–Wounan indigenous peoples have joined forces to oppose the completion of the highway project (CEALP 1996). Despite intense pressure from development interests, the presidents of Colombia and Panama signed a joint declaration agreeing not to complete the road. Questions remain, however, concerning the institutional relationship between Darien National Park and the Choco homeland, a politically autonomous unit of governance known as a Comarca (Clay 1988). These are the same issues and dilemmas that are found in every component of the corridor.

Experiments in bioregionalism: The Mesoamerican laboratory

Contemporary efforts to accommodate the conservation aspirations of local peoples purport to ignore, or at least subordinate, traditional jurisdictional boundaries, and foster mechanisms to cross political borders while moving responsibility for resource management to a more effective scale of governance – those who dwell in the land. As Doug Aberley described in Chapter 2, the integration of local, place-based culture with regional, ecological systems is one important bioregional value. Throughout the world, interesting experiments in institutional design have begun with grants of authority to local peoples, communities and indigenous groups. Central America has become a laboratory for many of these experiments in bioregionalism. Their effectiveness may ultimately determine the integrity of the Mesoamerican Biological Corridor.

Mesoamerica boasts one of the most successful efforts to accommodate indigenous self-determination aspirations within the constraints of the modern nation-state. The Kuna Indians are among the most celebrated indigenous groups in the world (Herlihy 1989). Having established the first autonomous homeland in Central America, and the first indigenous forest park in the world, the Kuna are examples of how conservation, indigenous and nation-state development objectives can coincide with exemplary results. A prominent Kuna lawyer once noted: "[we] aren't conservationists; rather [we] know how to relate humans and nature. This is the basic principle of indigenous people" (Gonzalez 1992).

At first glance, the Kuna Comarca and its unique forest park would appear to have resolved conservation's conundrum in the context of the modern nation-state. The Kuna have maintained their political autonomy: they control their land and resources. Moreover, the Kuna embody a bioregional sensibility. The Kuna Comarca occupies a 124-mile-long corridor of Panamanian Atlantic Coast rain forest and associated offshore islands. The approximately 40–50,000 Kuna live primarily on the hundreds of offshore islands while revering the mainland forest as a sacred place. When an overland link to the Comarca was built in the early 1970s, the mainland forest was threatened by invasion and consequent deforestation. In response, the Kuna developed PEMASKY, a Spanish acronym for the Kuna Wildlands Project, and secured donor funding to develop the Kuna Park. A key concept behind the park is to integrate Kuna traditional knowledge with Western science to create a "new synthesis" (Chapin 1993).

However, global forces are eroding the Kuna's bioregional value-system. Young Kuna are increasingly moving to urban centers and adopting Western values, inhibiting the intergenerational exchange of traditional knowledge. Efforts to achieve the "synthesis" of traditional knowledge and western science in Kuna Park have apparently suffered as a result. Commentators have expressed concern that the erosion in traditional knowledge will change the way the Kuna relate to nature.

In addition, many commentators believe that the factors that have brought conservation success to the Kuna may limit its transferability to other political and cultural contexts (Chapin 1993). The Kuna have lived on offshore islands since the middle of the last century, drawing their subsistence from a bountiful sea. This has given them the "luxury" of reserving the interior forest as a sacred site in their cosmology. The Kuna have an effective system of political governance which has been adapted to Western political systems.

There are other examples of protecting both cultural heritage and biodiversity. Indigenous rights advocates and conservationists in the Mosquitia Corridor have begun utilizing two powerful empowerment tools in an effort to protect the integrity of the Mosquitia biocultural region. Participatory research mapping and zoning is an ambitious effort to involve indigenous peoples in the Western traditions of cartography and land-use planning, while at the same time reflecting indigenous knowledge in the management application of these tools. Participatory research mapping uses local knowledge specialists and indigenous surveyors to map settlement distributions and subsistence land-use areas to produce cartographic data on indigenous lands. This methodology introduces indigenous peoples to Western conceptions of boundary definition while

providing indigenous peoples and government officials with information necessary to process land-claims. It also identifies areas of resource-use conflict
between groups and villages. In 1992, Herlihy (1996) and others employed this
technique in Mosquitia and produced the first settlement and land-use maps for
local organizations from the Miskito, Garifuna, Tawahka Sumu and Pesch
indigenous groups. The researchers contend that these maps have already
served to forestall commercial forest concessions on indigenous homelands.

In hope of advancing this effort, researchers have proposed "participatory
research zoning" – a method used to integrate indigenous land-use practices
with protected area conservation goals. Participatory research zoning is being
used in the Rio Platano and Tawahka Asangni Biosphere Reserves in the
Mosquitia. The method initially employs a "cognitive land-use mapping"
approach to establish the cultural ecology of the homelands. Researchers and
local knowledge specialists then classify and map indigenous land-use practices
and traditions. This is significant for management since in the case of the
Tawahka Asangni Biosphere approximately 95 percent of the reserve is used for
indigenous hunting, fishing and subsistence forest extraction, while the
remaining 5 percent is devoted to agriculture. With this classification and map
of land-uses in mind, indigenous surveyors verify the information and seek to
establish consensus concerning the classification scheme and maps. Researchers
work with community leaders to standardize the classification scheme among
the different groups and reconcile them with traditional conservation-zoning
classifications. A second consensus-based exercise ensures the integrity of the
second zoning iteration and seeks to resolve any conflicts over use boundaries.
The final step integrates the maps into a geographic information systems (GIS)-
compatible format, and incorporates the maps and classification system into
reserve management plans.

At the same time these "soft empowerment techniques" are being employed,
indigenous groups and their advocates are seeking to establish unfettered rights
to land and resources in the Mosquitia by using conventional Western land-
titling techniques. Indigenous groups recognize that Western conceptions of
property rights predominate in a world of nation-states, and that legal title
forms the basis for the tenure security they seek. However, commentators have
pointed out that in Nicaragua legal title by itself is insufficient to guarantee
tenure (Hendrix 1992). In addition, Western tenure is often ill-adapted to
indigenous customary law and land-use practices.

In the Nicaraguan Mosquitia, attorneys for indigenous groups and their
advocates are seeking to establish the legal basis for communal land-title
(Jarquin 1993), and to document community land-claims through the use of
participatory mapping techniques, ethnographic studies and conventional
records (Anaya and Crider 1996; McDonald 1996). This process has been
greatly complicated by resettlement due to the civil war, invasions by
nonindigenous settlers, the unstable political and administrative setting of the
Nicaraguan Atlantic Coast, and conflicts among groups and subgroups over use
areas.

Similar efforts are being undertaken in the Honduran Mosquitia (Wilbur 1996). Prominent conservation groups are supporting and underwriting these efforts, which they view as the best mechanism to achieve their biodiversity conservation goals in a manner that is compatible with indigenous self-determination aspirations (Stocks 1995; Anaya and Crider 1996; Wilbur 1996).

In addition to a number of these legal and consensus-based strategies, community-based resource management has been developed to deal with a number of conflicts. For example, lands in the multiple-use zone of Guatemala's Maya Biosphere Reserve are national lands under the control of the government of Guatemala. Originally inhabited by the ancient Maya civilization, in modern history the Guatemala Petén has been a forest frontier occupied by a hardy band of nonindigenous activists known as "peteneros." In addition to the so-called "minor forest products" like chicle and xate, the Maya Forest still harbors significant stands of commercially valuable hardwoods, potentially vast petroleum reserves and growing opportunities for tourism development. The government retains control of these resources. Since the reserve was established in 1989, there has been a moratorium on commercial forestry as the government sought to establish an effective regulatory framework for forestry in the reserve. In fact, the government has left an institutional vacuum. In the interim, illegal logging continues, with much of it moving into Mexico and Belize through border protected areas. At the same time, the region has become the final frontier for colonization. Landless peasants have been streaming into the region, illegally occupying the reserve, and practicing unsustainable "slash and burn" agriculture. Even among the nontimber forest product harvesting activities, there is growing concern that the resources are being unsustainably exploited by a new generation of activists who don't share their predecessors' traditional resource stewardship values and techniques. Governments have been powerless to change the over-exploitation of natural resources.

To deal with these problems, the Guatemalan government agreed to permit several small Petén communities within the reserve to seek permission to communally manage their "homelands" through integrated resource concessions. The communities must develop management plans and sign concession agreements with the government. The communities may engage in commercial forestry, ecotourism and other activities that are consistent with the reserve's purposes. Proponents believe that these communities have the necessary stake in the resources under their control to most effectively protect and sustainably manage them, and are poised to provide technical assistance. However, this effort at community-based natural resource management is confounded by several factors that are predicates for bioregional management. The influx of new residents to the region has diluted the community cohesiveness that would appear to be a requisite for community management. These communities lack the degree of cultural identity and intergenerational commitment typically found in indigenous communities. Moreover, it is questionable whether the geographic boundaries of the concessions adequately coincide with the distribution and range of the wide-ranging "chiclero" and "xatero" activists whom conservationists hope will be primary beneficiaries of the new arrangement. Instilling a sense of community identity and custodianship will require special attention. In addition, the community will have to seek special institutional arrangements with lands outside of the concessions to ensure the viability of their extractive activities.

The lens of the ancient Maya: Advancing bioregional policy

Joined by an ancient causeway that straddles the political boundary between Belize and Guatemala, the ancient Maya center of El Pilar is perhaps the clearest example of bioregionalism. Although the two countries enjoy cordial relations, long-standing territorial disputes have tempered efforts to achieve full bilateral cooperation. The portion of El Pilar in Belize comprises less than 1,000 acres. The land surrounding the proposed reserve (and some within) is occupied by small agrarian "milpa" farmers and a rapidly expanding tourism economy in nearby San Ignacio. The portion of El Pilar in Guatemala lies within the multiple-use zone of the Maya Biosphere Reserve. This portion is relatively undeveloped and largely forested, in considerable contrast to the Belizian context. Only recently discovered, no detailed effort has been made to map the extent of the site.

In an effort to establish a reserve unique in its bi-national nature, researchers at the site have initiated an effort at community-based bilateral cooperation between the two nations (Ford and Montes 1996). Archaeologists, anthropologists and restorationists working at El Pilar are developing a strategy that will allow for the management of the Maya Forest through "the lens of ancient Maya." The Maya inhabited the Petén Forest in numbers estimated to be as great as ten times the present population of the region, presumably extracting resources in a sustainable fashion for centuries until the civilization's eventual collapse. El Pilar's managers are attempting to interpret the ancient Maya lifestyle by developing a modern Maya "forest garden," based on evidence of ancient Maya polycultural farming practices. Local villagers with traditional knowledge are assisting in this effort. Integrating El Pilar into the fabric of the broader community is a stated management objective. Additionally, site researchers favor an interpretation scheme that emphasizes the way of life of the nonelite among the Maya.

Site researchers are attempting to conserve and restore a remnant of the Maya Forest, and the storehouse of tropical biological diversity that *grew over* an ancient civilization. Protected areas within the Maya Biosphere Reserve serve as refugia for the flora and fauna that characterize this forest. Recent attention has been on the role of these refugia in a larger land-use mosaic that can sustain viable populations of the flora and fauna of the Maya Forest. El Pilar's transboundary nature provides a strategic linkage within the larger tri-national mosaic of the forest. Efforts to link El Pilar to sustainable-development initiatives outside the proposed reserve can extend its influence within this larger mosaic.

The El Pilar mandate to share management of a single resource located in two countries presents what is the ultimate challenge for bioregionalism – accommodating separate political sovereignties that share a common resource. Bioregionalism will require a framework that can accommodate the legal and administrative requirements of the separate sovereignties involved as seamlessly as possible. Such a framework would provide for one management plan implemented by two management units, each representing the portion of the resources located within each country – El Pilar in Belize and Pilar Poniente in Guatemala (Ankersen, Montes, Balderamos and Ortiz (forthcoming) 1997). Figure 10.2 provides an institutional map for a proposed bi-national comanagement arrangement.

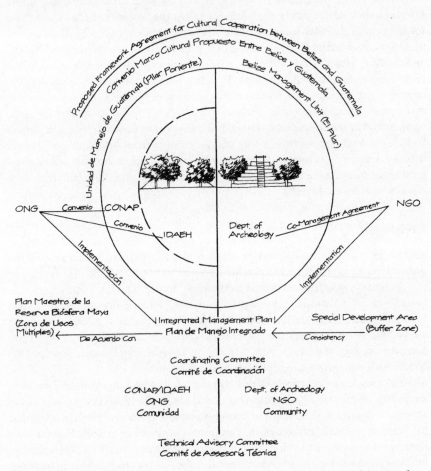

Figure 10.2 Proposed institutional framework for the El Pilar Archeological Reserve for Maya Flora and Fauna

Source: El Pilar Roundtable, Mexico City (19–24 January 1997)

Figure 10.2 depicts in graphic form an institutional framework for El Pilar proposed by the research group at a 1997 roundtable to discuss the reserve's development in Mexico City. Both Guatemala and Belize authorize the management of protected areas by non-governmental organizations through concessions or comanagement agreements. Researchers hope that community-based non-governmental organizations from each country will assume the responsibility for management of the respective management units.

To address management issues common to both units, a coordinating committee comprising the appropriate representatives from governmental resource agencies, non-governmental entities and community-members who are involved in management should be established to ensure coordination and consistency with the agreed management plan by each management unit. Overarching this management framework should be a broader bilateral effort for cultural exchange between the two nations, that could include mechanisms for the resolution of management conflicts at the site.

Despite the literature concerning the establishment of "bi-national parks" and "peace parks," research has revealed no instances where established protected areas are truly integrated across national borders. Achieving the objective of the El Pilar Reserve may represent an innovation in reserve management, and a significant advance in bioregional policy. Nonetheless, there remain significant issues to be resolved relating to the nature of the delegations to non-governmental organizations, the forms of international agreements that may be required, equitable financial mechanisms, the nature of community participation, site security and immigration concerns; and appropriate mechanisms to resolve disputes.

Conclusion

The conservation conundrum in the planning and management of the Mesoamerican corridor shows that governments continue to pay lip service to the devolution of resource-control to the most appropriate local and ecosystem-based scale, but appear unwilling to take the "bioregional plunge" toward fostering local, community-based self-determination. Even where land-rights are acknowledged, governments remain reluctant to relinquish control of natural resources within reserves. Similarly conservationists continue to pay lip service to indigenous aspirations for self-determination.

The case studies described in this chapter reflect the complexities inherent in designing instruments and institutions to sustain cultural and ecological diversity. Each case study offers valuable lessons to bioregionalists who are interested in preserving local knowledge and culture, maintaining native diversity and fostering self-determination. The challenges facing the Kuna Yala experiment in indigenous autonomy may temper bioregionalism's exaltation of the indigenous lifestyle. Globalization and colonialization are major barriers to protecting indigenous autonomy. The "Westernization" of indigenous societies remains a barrier to cultural sustainability. In the Mosquitia, in their efforts to regain a measure of sovereignty, indigenous groups and their advocates are adopting – or co-opting – the traditional tools of Western land subdivision, titling, and zoning. In the Maya Forest, conservationists are placing their faith in labor-intensive low technology for chewing gum and ornamental plants. At the same time, in El Pilar, researchers are extolling the sustainability model of the ancient Maya, a civilization that some believe collapsed from its own weight on the land. The success of these and other similar efforts will determine the extent to which the conservation conundrum can be resolved in Mesoamerica, and the extent to which one of the most profound bioregional phenomena in the world, the Mesoamerican Biological Corridor, can be conserved.

Bibliography

Anaya, J. and Castellón, M.L.A. (1995) *Petition by the Mayagna Indian Community of Awas Tingi and Jaime Castillo Felipe against Nicaragua*, Inter-American Commission on Human Rights, Organization of American States.

Anaya, J. and Crider, A. (1996) "Indigenous People, the Environment and Commercial Forestry in Developing Countries: The Case of the Awas Tingi, Nicaragua," *Human Rights Quarterly* 18 (2): 345–67.

Ankersen, T.T. (1994) "The Mesoamerican Biological Corridor: The Legal Framework for an Integrated Regional System of Protected Areas," *Journal of Environmental Law and Litigation* 9: 499–547.

Ankersen, T.T. and Trujillo, H.G. (1996) "Confronting the Crisis: Conservation Law and Policy in the Maya Forest," *Vida Silvestre Neotropical* 4 (2): 85–8.

Ankersen, T.T., Montes, J.A., Balderamos, D. and Ortiz, A. (1997, forthcoming) "Advancing Bioregional Policy: Integrated Management of a Shared Cultural and Natural Resource – The Case of the El Pilar Reserve for Maya Flora and Fauna," Washington, D.C.: World Monument Fund.

Berkey, L. (1994) "Maya Land Rights in Belize and the History of Indian Reservations: A Report to the Toledo Maya Cultural Council," presented to the Indian Law Resource Center, Washington D.C.

CEALP (Center for Popular Legal Assistance) (1996) *Roads Through the Rain Forest: Indigenous Peoples Battle to Preserve the Darien Gap* (Winter 1996), Report no.2.

Chapin, M. (1993) "Recapturing the Old Ways: Traditional Knowledge and Western Science among the Kuna Indians of Panama," in C.D. Kleymeyer (ed.) *Cultural Expression and Grass Roots Development*, Boulder and London: Lynne Rienner Publishers.

Clay, J. (1988) "Conservation and Indigenous Peoples: A Study of Convergent Interests," in J. Bodley (ed.) *Tribal Peoples and Development Issues: A Global Overview*, Mountain View: Mayfield Press.

Craighead, F. (1979) *Track of the Grizzly*, San Francisco.

Ford, A. and Montes, J.A. (1996) "Environment, Land Use and Sustainability: Implementation of the El Pilar Archaeological Reserve for Maya Flora and Fauna, Belize-Guatemala," unpublished paper submitted to the International Conference on Land Tenure, Orlando, Florida.

Gonzalez, J. (1992) "We are not Conservationists," *Cultural Survival Quarterly* Fall: 43–5.

Grimes, A. (1993) "Indigenous Land Titling to Support Biodiversity Conservation in the Bosawas Natural Resource Reserve," Workshop (December 13–1), Managua: General Observations USAID LAC TECH Project.

Guillen-Trujillo, H. (1995) "Land Tenure and Conservation Conflicts in the Lacandon Forest, Chiapas, Mexico," *Land Tenure and Conservation Conflicts*, Mesoamerican Environmental Law Program Publication Series no. 2, Gainesville: University of Florida.

Hendrix, S. (1992) "The Crisis in Land Law and Policy in Nicaragua," *Comparative Juridical Review* 29, Coral Gables: Rainforth Foundation.

Herlihy, P. (1989) "Panama's Quiet Revolution: Comarca Homelands and Indian Rights," *Cultural Survival Quarterly* 13 (3): 17–24.

—— (1990) "Wildlands Conservation in Central America during the 1980s: A Geographical Perspective," *Benchmark* 17–8: 31–43.

—— (1996) "Indigenous Peoples and Biosphere Reserves: Culture and Conservation of the Mosquitia Corridor, Honduras," in S.F. Stevens (ed.) *Conservation through Cultural Survival*, Washington D.C.: Island Press.

IUCN (The World Conservation Union) (1994) Guidelines for Protected Area Management Categories, Washington, D.C.

Jarquin, L. (1993) "Diagnostico Legal Acerca de la Titulacion de las Comunidades Indigenas Ubicadas en la Reserva Bosawas," paper on file with author.

Korten, A. (1994) "Closing the Darien Gap? The Pan-American Highway's Last Link," *Abya Yala News* 8 (1–2): 29–30.

McDonald, T. (1996) "Awas Tingi: Un Estudio Etnografico de la Comunidad y su Territorio: Informe Preliminar," unpublished manuscript on file.

Meyers, N. (1996) *Ultimate Security: The Environmental Basis of Political Stability*, Covello: Island Press.

Nations, J.D. (1988) "The Lacandon Maya," in J.S. Denslow and C. Padoch (eds.) *People of the Tropical Rain Forest*, Berkeley: University of California Press.

Neitschmann, B. (1990) "Conservation and Conflict in Nicaragua," *Natural History* November: 42–9.

—— (1995) "Conservation, Self-determination and the Miskito Coast Protected Area, Nicaragua," *Mesoamerica* 16 (29): 1–157.

Redford, K. (1991) "The Ecologically Noble Savage," *Cultural Survival Quarterly* 15 (1): 46–8.

Ross, J.F. (1996) "A Few Miles of Land Arose From the Sea," *Smithsonian* 27 (9): 112–21.

Sader, S. (1996) *Forest Monitoring and Satellite Change Detection Analysis of the Maya Biosphere Reserve, Petén District, Guatemala*, final report submitted to Conservation International and US Agency for International Development.

Sale, K. (1991) *Dwellers in the Land: The Bioregional Vision*, San Francisco: Sierra Club Books.

Schwartz, N. (1990) *Forest Society: A Social History of Petén, Guatemala*, Philadelphia: University of Pennsylvania Press.

Stocks, A. (1995) "Land Tenure, Conservation and Native Peoples: Critical Development Issues in Nicaragua," unpublished paper presented at the Society for Applied Anthropology Meetings.

Stone, M. (1995) "The Cultural Politics of Maya Identity in Belize," *Mesoamerica* 16 (29): 167–214.

Wilbur, S. (1996) "The Honduran Mosquitia: A Pre-Investment Analysis for the Parks in Peril Program," Arlington: The Nature Conservancy.

World Resources Institute (1997) *Key Characteristics of Bioregional Management*, <http://www.wri.org/wri/biodiv/b014-bts.html>

Part IV

Toward a bioregional future

11 The role of education and ideology in the transition from a modern to a more bioregionally-oriented culture

Chet A. Bowers

As we learn more about the extent of the ecological crisis, it becomes easier to recognize that the forms of knowledge privileged in our public schools and universities are not contributing to the enlightenment and general progress of humankind – this continues to be the litany of educators ranging from elementary teachers to university professors and presidents. Rather, the high-status forms of knowledge promoted through our educational institutions (particularly our most prestigious universities) contribute to the disintegration of previously self-reliant cultural groups, to widespread chemical changes in the life-processes of the earth's ecosystems, and to the development of technologies and centralized systems of control that further degrade natural habitats already under stress. Moreover, the growing dominance of the high-status knowledge now equated with modernity in Third World countries can be traced directly to the Western form of education that their elite classes have been exposed to. Why are the connections between the globalization of Western forms of knowledge and the ecological crisis not more widely recognized within the public school and university communities? The answer can, in part, be found in the cultural ideology (which can also be understood as a culturally specific epistemology) that is encoded and reproduced in the language of these institutions, which serves both as the basis of knowledge in the various disciplines, and as giving moral legitimization to efforts to replace the traditional forms of knowledge that coevolved within bioregions with modern ways of understanding and technical expertise. The nature of this guiding and legitimating ideology, as well as why it should be abandoned in favor of the conceptual and moral orientation of cultural bioconservatism, will be the main focus of this chapter. Examining the connections between high-status knowledge, the globalization of hyper-consumerism and the hyper-technologizing of relationships, as well as the connections between cultural bioconservatism and the renewing of ecologically centered cultural practices, represents another way of framing the educational issues that should be part of any discussion of bioregionalism.

In 1971, Ivan Illich's *Deschooling Society* appeared as yet another attack on the educational establishment. But even among radical educators who were attracted to the cogency of his arguments, Illich seemed overly romantic (even reactionary) and politically naive compared to the seemingly more revolutionary

192 Chet A. Bowers

possibilities of a Marxist analysis of how schools contribute to the reproduction of class relationships. While Illich did not focus primarily on how Western forms of education contribute to the ecological crisis, his arguments read today as nearly identical to Vandana Shiva's analysis (1993) of the spread of monoculture, and to the arguments about the colonizing effects of Western sponsored development in Wolfgang Sachs' two edited collections of essays (1992; 1993). According to Illich:

> Obligatory schooling inevitably polarizes a society: it also grades the nations of the world according to an international caste system. Countries are rated like castes whose educational dignity is determined by the average years of schooling of its citizens, a rating which is closely related to per capita gross national product. . . . The paradox of the schools is evident: increased expenditure escalates their destructiveness at home and abroad.
>
> (Illich, quoted Sachs 1992: 9)

Later in the book, Illich pinpoints the nature of this destructiveness, as well as what he thought was the sustainable pathway that the cultures of the world would need to follow:

> I believe that a desirable future depends on our deliberately choosing a life of action over a life of consumption, on our engendering a life style which will enable us to be spontaneous, independent, yet related to each other, rather than maintaining a life style which only allows us to make and unmake, produce and consume – a style of life which is merely a way station on the road to the depletion and pollution of the environment.
>
> (1992: 52)

Illich's criticisms of how Western approaches to education contribute to the commoditization of knowledge and relationships, as well as an increased dependency upon environmentally destructive technologies, quickly disappeared from the educational scene. While radical educators were taking the long march from structural Marxism, through critical theory, and now into the equally anthropocentric discourses of postmodernism and cultural studies, criticism from within the educational establishment shifted from the "Nation at Risk" report to the debate over preserving the Western canon and fostering cultural literacy (basic factual knowledge), and is now focused on how to incorporate computers into all phases of the educational process – which, according to Sherry Turkle, enables "students to construct new selves through social interaction" in cyberspace (1996: 151).

As we learn more about the ecological consequences of Western-based technologies, which are dependent upon the forms of knowledge learned in Western-style universities, the question of why only a few environmentalists and even fewer academics have recognized the relationships that Illich saw so clearly grows more significant. The answer that is the most compelling for me is that the ideology acquired in their public schools and university education shaped the deepest and most taken-for-granted symbolic foundations of their thinking in such a way that even environmental setbacks are interpreted as part of the human-centered narrative of progress. For example, how many scientists working on the development of new synthetic chemicals will really think seriously about the connections between their research and social progress after having read about the widespread introduction of hormone-disrupting chemicals into the environment (Colborn, Dumanoski and Myers 1996)? And how many economists will question the wisdom of including environmentally destructive practices in their measurement of the nation's economic growth? Indeed, the myth of progress continues to be such a taken-for-granted basis of interpretation that what should lead to a radical rethinking of basic assumptions is too often viewed as a momentary setback with no long-term implications. As I have written elsewhere about the connections between ideology, language and the taken-for-granted patterns of consciousness (Bowers 1993a; 1993b; 1995; 1997), I will summarize the key elements of the ideology that serves as the basis for separating knowledge into the categories of high- and low-status, and for perpetuating in only slightly altered form the deep assumptions that were the basis of the Industrial Revolution. To make the last point in a somewhat different way, the shift from a reliance on production based on cheap human labor and heavy machinery to digital computers has not been accompanied by fundamental changes in the ideology that coevolved with the Industrial Revolution.

Characteristics of the ideology that underlies high-status knowledge

The deep cultural assumptions that underlie the ideology that was used to legitimate the Industrial Revolution and now legitimates the Information Age, can be seen in Hans Moravec's optimistic projection of the future of human evolution. Moravec's view is similar to the other globalists. In a book published by Harvard University, *Mind Children: The Future of Robots and Human Intelligence*, Moravec describes the transition that human culture is undergoing:

> Our culture still depends utterly on biological human beings, but with each passing year our machines, a major product of the culture, assume a greater role in its maintenance and continued growth. Sooner or later our machines will become knowledgeable enough to handle their own maintenance, reproduction and self-improvement without help. When this happens, the new genetic takeover will be complete. Our culture will then be able to evolve independently of human biology and its limitations, passing instead directly from generation to generation of ever more capable intelligent machines. . . . A postbiological world dominated by self-improving, thinking machines would be as different from our own world of

living things as this world is different from the lifeless chemistry that preceded it.

(1988: 4–5)

While many academics would probably challenge Moravec's statement about the "postbiological world," they nevertheless share the deep cultural assumptions that underly his futuristic and global vision. These assumptions include:

Autonomous individualism   Moravec represents himself as a rational individual who is giving an objective description of changes occurring in the transfer of human intelligence to machines. That is, he takes for granted the earlier arguments of John Locke and René Descartes that the properly grounded thought process of individuals is free from the influence of their culture's epistemological traditions. This view of the individual (as organizing ideas on the basis of external data/experience) leads, in turn, to representing knowledge as objective and needing only to meet the criteria that governs what Alvin Gouldner referred to as the "culture of critical discourse" (1979), which will be discussed later in the chapter.

Progressive nature of change   Moravec projects the existential stance of the objective scientist who is simply explaining the dynamics of the evolutionary process. But his inability to acknowledge the vast amount of evidence that machine-based progress has undermined the viability of cultures and natural systems, and contributed to huge disparities in the distribution of wealth within and between cultures, is directly related to how unconsciously held assumptions influence what is recognized and what is ignored. His ability only to see the connections between humans and computers, his inability to recognize the destructive character and use of machines, and the interpretation that he presents, are only possible because his culturally learned schemata equate change with a linear form of progress.

Anthropocentrism   Although Moravec's statement explains how human genetic coding is going to be replaced by a machine-based form of intelligent encoding, it still retains a deeply anthropocentric way of thinking. Humans are represented as being the only intelligent form of life; indeed, the only form of life capable of "downloading" their intelligence into intelligent machines that will be free of the limitations of human mortality.

Authority of scientific knowledge/empowerment through technology   The explanatory framework (cosmology) of the evolutionary process is basic to Moravec's account of human/machine relationships over time, as is the assumption that science-based explanations are the only ones that can be legitimately considered by rational individuals. The reductionist way of thinking in today's high-status areas of scientific research and cosmology relies increasingly on machine-derived metaphors as the basis of understanding, and provides the knowledge necessary for the development of biotechnology and artificial intelligence – which are the two most important areas of technological development driving the process of globalization.

Commoditization of the commons   While Morovec's statement does not directly argue for the further commoditization of knowledge and relationships, the "intelligent machine" he views as the next stage in evolutionary development is part of the market consumer-oriented culture that is now displacing the norms of reciprocity that characterized the commons in tribal and traditional cultures. That is, computer-mediated thought and communication require that human beings participate in the market economy in ways that face-to-face based cultures do not.

The deep culturally specific assumptions about the autonomous nature of the individual (in all modes of expression – including language, thought, creativity, moral judgment, etc.), the progressive nature of change, the anthropocentric view of nature, the increasing authority being claimed on behalf of scientific explanations, the reliance on technological approaches to the redesign of nature, and the commoditization of knowledge and relationships, should not be seen simply as the conceptual and moral basis of the Industrial Revolution and now the Information Age. They also serve as the basis for determining the difference between high- and low-status forms of knowledge, and thus what will be learned in our public schools and universities. To put this another way, these assumptions underlie the forms of knowledge deemed essential to the educated and modern individual. Graduates of Western universities, for example, become part of the emissary tradition of globalizing these assumptions in the name of modernization and globalization – which they view as having the same historical inevitability that is communicated in Moravec's description of our transition to a "postbiological world."

These essential elements of the ideological /epistemological foundations of modern, high-status culture are shared across academic disciplines that seemingly address different issues and require different ways of thinking. They are basic to the different interpretations of educational liberalism that public school teachers and university professors identify themselves with. The technologically-oriented educators promoting computer-based learning and a systems approach to instruction simply foreground different assumptions from those that the emancipatory and neo-Romantic educators regard as their primary concern. For example, technocratic liberals are more explicit about the importance of relying on scientific and technologically-based experts for improving education. While this orientation leads them to ignore the emancipatory liberal's language of individual empowerment, and the focus on the natural goodness and efficacy of self-initiated learning that is the primary concern of neo-Romantic educators, they nevertheless share with these other liberal approaches to education (and with scientists, economists, philosophers, historians and so on at the university level) the same assumptions about individualism, progress, anthropocentrism and so forth. The extent to which these modern assumptions are shared by seemingly different traditions can be seen in how they are now being embraced by advocates of the virtual classroom, early-childhood educators and professors who identify themselves as postmodern thinkers.

The assumptions of the Industrial Revolution/Information Age can be seen in the rules of critical discourse that are represented throughout the university community as the basis of legitimate knowledge and of progress itself. These rules serve as the basis for dismissing other forms of knowledge as low-status and thus inferior. According to Alvin Gouldner (1979: 28–9), the form of discourse viewed within the university community as essential to the advancement and universalization of high-status knowledge includes the following: (1) knowledge statements must be justified; (2) the mode of justification must be based either on empirical evidence or the power of current theory and not on the authority of tradition, sacred texts, or the social position of the speaker; (3) the process of justification requires that the participants be free to reach their own conclusions about the merits of the evidence or theoretical justifications.

While the rules of critical discourse appear not to set limitations on what can be relativized through the critical scrutiny and argument, the rules themselves are based on the same assumptions as Adam Smith's "invisible hand" that ensured that in the free play of market forces, the most deserving economic decisions will prevail. The bias against tacit, contextual and traditionally grounded forms of knowledge, as well as the emphasis on the ability to utilize theory and systematically organized evidence, ensures that the "marketplace" of ideas will be dominated by those individuals who possess the form of communicative competence favored by the rule of critical discourse. By making the central purpose of serious and legitimate discourse the overturning of traditional forms of authority (which is magnified by the pressure on young faculty members to publish in order to be promoted and tenured), universities contribute to the knowledge explosion which is seen as a sign of progress. That it is mostly highly abstract theoretical knowledge untested in the life of a community, that it will be replaced by even newer ideas and technologies before its full meaning and usefulness can be explored, and that it generally undermines the accumulated wisdom about relationships within a community and its bioregion, are not seen as diminishing the contribution of this knowledge to the modernizing and globalizing process.

In addition to privileging people who learn the rules of critical discourse and possess the elaborated speech-codes of the various disciplines, universities contribute to the grading of cultures in terms of their possession of high-status knowledge by emphasizing the authority of the printed word. Indeed, the degree of literacy within a culture has been viewed as a hallmark of modern, developed cultures; and the lack of literacy has been one of the main criteria for judging a culture to be "undeveloped," "backward" and "primitive." The current scholarly interest in studying the differences between orality and literacy, as well as the growing acceptance of research methodologies based on narrativized forms of knowledge, have not fundamentally altered the privileged status that universities accord to print-based thinking and communication – a tradition that is being strengthened through the increasing use of computers in classrooms and scholarly research.

While literacy has important uses, given the presence of other cultural assumptions it contributes to the culturally and ecologically destructive characteristics of high-status knowledge. When it is the privileged mode of encoding knowledge within a culture that is based on deep assumptions about the autonomous individual, anthropocentrism, progress and so forth, literacy helps strengthen the taken-for-granted patterns of abstract, analytical and individualistic thought. Fixed texts that encode knowledge separated from context and that hide the cultural language/existential processes connected with writing (articles, books, software programs), reinforce an individualistic way of knowing. But the form of consciousness being reinforced has implications that go beyond strengthening the authority of subjective interpretation; literacy also contributes to the modern goal of education as contributing to the mobility of the individual. What is learned is highly abstract and seldom useful in terms of understanding the patterns of different systems that make up a bioregion; literacy-based knowledge is represented as contributing to citizenship within the global community. Unfortunately, this vision of the global community, which is now being promoted by economic and technological elite groups, does not take into account the double-bind of promoting the expansion of the Western consumer lifestyle within the context of ecosystems that are in rapid decline.

Low-status knowledge, cultural bio-conservatism and bioregionalism

Low-status knowledge has no legitimate standing within the academic community of critical discourse, unless it is made the subject of analysis and debate. It tends to be communicated more on a face-to-face basis, rather than through mechanical or electronic print. It has more to do with decisions influenced by the needs of putting food on the table, helping neighbors, knowing when and where to plant, entertaining guests and celebrating a holiday, observing the appropriate reciprocal norms that are at the center of a community ceremony, knowing where to gather medicinal plants, being able to design a building that uses local materials and utilizes the sources of energy from the environment, and so forth. In short, low-status knowledge encompasses the knowledge accumulated over generations of communal experience with the cycles and patterns of life-forms that make up the environment. Professors of folklore and some anthropologists study this form of knowledge, but seldom for the purpose of seeking ways of restoring the vitality of the commons that continue to be subverted by the forces of commoditization. While the ecological crisis is becoming an important field of research for a small number of academics, their continued adherence to the rules governing the culture of critical discourse too often results in their scholarly efforts adding more books to university libraries, and spending time on the internet sharing their data with scholars in other parts of the world. Their efforts may lead to important environmental restoration projects, changes in environmental legislation, and a more in-depth under-standing of changes affecting the environment; but they seldom contribute to the local knowledge of cultural groups that are attempting to avoid being drawn into the cycle of global development, modernization and commoditiza-tion of life patterns that previously were sustained through complex traditions of intergenerational sharing.

Any discussion of the connections between low-status knowledge, viable face-to-face communities and ecological sustainability needs to avoid the mistake of associating low-status knowledge only with ecologically-centered cultures. Bioregionalists should recognize that an increasing percentage of the world's population live in cities and that one of the most daunting challenges will be to help citified populations evolve less consumer- and energy-dependent forms of existence. One aspect of this challenge will be to identify attenuated forms of low-status knowledge that sustain face-to-face relationships – in activi-ties ranging from quilting and singing groups, chess clubs, sports associations, to urban gardening – and thus the experience of belonging to a community of reciprocal relationships and responsibilities. Whatever strengthens this experi-ence of community-participation helps to provide a point of contrast to the depersonalized forces of computer-based technology and consumerism. Later in the chapter, I will discuss how these face-to-face, noncommoditized relation-ships that still exist in ethnic cultures, and even within the dominant culture that has moved to the suburbs, need to be incorporated into the curriculums of public schools and universities as examples of sustainable cultural patterns.

Before addressing the educational issues related to the recovery of more self-reliant community bioregional relationships, I want to explain why cultural bioconservatism must replace the various expressions of liberalism as the guiding ideology (and cultural epistemology) for thinking about the reform of education. To reiterate a point made at the beginning of the chapter, the various expressions of liberalism used to legitimate different approaches to educational reform are based on the deep cultural assumptions that were taken for granted by the elite groups responsible for the Industrial Revolution. The double-bind of representing as progressive the assumptions that are responsible for the Industrial Revolution suggests the need for a more accountable use of our political language: accountable in a way that represents the basic relation-ships between human and biotic communities, and not the mythic vision of material progress that we know is unsustainable.

A careful reading of environmentalists such as Kirkpatrick Sale, Charlene Spretnak, Wendell Berry, Wes Jackson, Vandana Shiva, Masanobu Fukuoka and Dolores LaChapelle does not reveal a thought-process based on liberal assump-tions about progress, an anthropocentric universe and the moral and intellectual authority of the autonomous individual. Rather, they are writing about the knowledge of place, the reciprocal norms that govern community life and the ways in which modern approaches to technology and material forms of progress degrade both self-reliant capacities of communities and the renewal of natural systems. In different voices, they are explaining how conserving and renewal are part of the same life process. And when we read Richard Nelson's description of the hunting practices of the Koyukon, J. Stephen Lansing's account of how the Balinese temple system is integrated into the ecological rhythm of rice farming, and Robert Lawlor's explanation of how the Australian Aboriginal cultures encoded both spiritual and local place-based knowledge into their Songlines, we can begin to understand how cultures develop the ability to *conserve* forms of knowledge refined over generations of learning from the patterns of the local environment.

Although they do not use the term, Sim Van Der Ryn and Stuart Cowan provide an eloquent summary of what I call cultural bioconservatism. In writing about ecological design principles, they note:

> Local knowledge is best learned through a steady process of accretion. The knowledge of the careful farmer or rancher, with his or her long experience of soil, crops, livestock and weather, is an irreplaceable design resource. So is the knowledge of a traditional earth builder, a craftsperson, fisherman, a bird watcher, or a rower. The collective memories of those who inhabit a place provide a powerful map of its constraints and possibilities. In a sense, ecological design is really just the unfolding of place through the hearts and minds of its inhabitants. It embraces the realization that needs to be met in the potentialities of the landscape and the skills already present in a community.
>
> (1996: 65)

The key elements that distinguish this form of conservatism from temperamental, economic, religious and philosophical conservatism include the awareness that the "cultural accretion" grows out of deep and lasting relationships with the natural life-support systems – soil, water, animals, weather and so forth. This is radically different from the various expressions of liberalism which equate progress with the freedom to introduce new ideas, values and technologies. Progress is equated with new beginnings and what can be judged on the basis of what seems relevant to the individual's sense of time rather than in terms of knowledge accumulated and refined over generations of experience. It is generally ignored that the liberals' efforts to break from what they view as the constraints of traditions actually involve introducing what are, in fact, experiments into cultural and natural systems – experiments that lead too often to unanticipated consequences that cannot be reversed.

The other distinguishing characteristic of cultural bioconservatism is the orientation toward conserving (and renewing) cultural practices that are the basis of an equitable and sustainable form of community. Economic conservatism (which is really the contemporary expression of classical liberalism) treats nature as an exploitable resource. Philosophical conservatism, while offering insights into the complexity of civic life, has a strong anthropocentric orientation. Cultural bioconservatism shifts the focus away from the individual as the basic social unit. Indeed, the term "culture" helps to highlight that individuals are nested in cultural systems, and cultural systems are nested in natural systems. This layered way of understanding fundamental relationships leads to greater awareness than what liberals refer to as individual intelligence. It recognizes that the cultural way of knowing, from its mythopoetic narratives to its use of technologies, has been shaped by the patterns and changes occurring in nature over a period of time that exceed what can be comprehended by the individual celebrated by liberal thinkers.

Educational implications

In order for public schools and universities to play a constructive role in the transition to a more bioregional existence, it will be necessary for them to undergo radical changes. Currently, these institutions are the only ones in society that systematically socialize students to bodies of knowledge that have been built upon the assumptions of the Industrial Revolution. While public schools socialize students to an increasingly strange mix of neo-Romantic and Cartesian ways of thinking, universities have helped to create new metaphors for advancing knowledge – metaphors that have made it more difficult to recognize the epistemological continuities with the pivotal shift in the direction of cultural development that began some 350 years ago. Deconstructionism, postmodernism, cultural studies, microeconomics, sociobiology, cognitive neuroscience and process philosophy are just a few of the many new areas represented by their adherents as being on the cutting edge of inquiry – and thus the latest expression of progress in humankind's efforts to explain the past in ways that enhance the ability to control the future. While this proliferation of areas of inquiry within universities involves radically different interpretive frameworks, they have the effect of giving a progressive appearance to cultural assumptions that continue to undermine the authority of local knowledge, the wisdom encoded in mythopoetic narratives, the experience of the sacred, the importance of elder knowledge, the reliance on appropriate technologies that reflect local custom and need, and the nonmaterial ways of interpreting wealth that cultures have developed.

The universities' current role in globalizing the modern beliefs and values that complement the new technologies contribute to the further commoditization of everyday life – a concern reflected in the writings of Serge Latouche (1992) and the contributors to *The Development Dictionary: A Guide to Knowledge as Power* (Sachs 1992). The spread of the market mentality, from remote villages to over-crowded cities, suggests one of the changes that needs to be made at each level of the educational process: Public schools and universities need to combine a study of ecologically sustainable forms of culture with a critique of how high-status ways of thinking (including high-status technologies like computers) are destroying the viability of local cultures and introducing changes into natural systems that exceed the ability of these cultures to adapt to them. As public schools and universities were established to promote the prevailing body of high-status knowledge that individuals could take with them to any region of the world, my suggestions for educational reform may sound totally naive. But developing a critical understanding of the deep cultural assumptions that continue to be used to legitimate the globalizing of Western high-status culture seems an achievable first step toward a return to a more bioregional pathway of cultural renewal. Students at both public school and university level should study a number of the essential elements of modern culture in a way that is framed by the ideology of cultural bioconservatism – which is the starting point for understanding the connections between ecologically sustainable culture patterns and the characteristics of a bioregion. These areas of study include understanding the cultural mediating characteristics of modern technologies and how they differ from traditional, more bioregionally-centered approaches to technological development. The critical study of the cultural non-neutrality of technology should also include how technology influences thought- and language-patterns, as well as the deep cultural assumptions that become encoded in the technology.

A second area of study that needs to be reframed in terms of a cultural bioconservative orientation – rather than in terms of a liberal orientation that gives blanket legitimization to all forms of change as expressions of progress – is the combination of high-status areas of inquiry (economic, political, scientific, technological) that are transforming human activities, local knowledge and even the genetic basis of life into commodities that are being manufactured in increasingly workerless factories. Careful consideration of the impact of the commoditization mania in terms of the loss of local knowledge and patterns of reciprocity essential to viable communities (including the importance of work within the community) would be one way to help students understand that the real challenge is to conserve local traditions that provide an alternative to consumerism – and the many forms of dependency that accompany it.

As most areas of public school and university curricula do not include the study of the deep cultural assumptions that underlie the high-status forms of knowledge that students are expected to base their personal and professional lives upon, educators should focus more attention on the influence of modern culture. Making explicit the assumptions that influence, for example, how history is written and interpreted, would enable students to recognize the cultural basis for assuming that the technologies, ideas and values associated with modernization should be globalized. This would contribute to the dereification of the idea that modernization is inevitable and globalization is a reflection of the progress of modernity. The study of culture should include the language–thought connection – including how the root metaphors and mythopoetic narratives are an outgrowth of local experience and the danger that arises when these root metaphors and mythopoetic narratives are imposed on other cultural groups.

Each level of formal education should help students understand the characteristics of ecologically sustainable cultures. The following questions would help to focus attention on the cultural patterns that have been largely ignored because of the low-status they have been given by liberal educators:

What are the forms of activities (ceremonies, mutual aid, leisure, norms of
   reciprocity, etc.) that are prominent in traditional, ecologically centered
   cultures?
What are the forms of activities in modern urban (and suburban) communities
   that enhance the quality of life and do not depend upon consumerism?
What are the forms of association (sports, ceremonies and celebrations, mutual
   interest groups, civic activities, etc.) within the student's community that
   strengthen the sense of solidarity and reciprocity, and do not have an
   adverse impact on ecology?

As most of our cultural knowledge is learned at the tacit level, what is not
named and thus made explicit is further marginalized in our awareness by new
consumer items and other modern fashions given high visibility in the media. If
the educational process does not focus attention on the non-consumer-oriented
patterns of community life, it is less likely that students will realize (and value)
what is being lost through the relentless pressure to extend the market-
mentality into every area of life. The commoditization of access to an
increasingly wide array of information requires owning a computer and having
the financial means to continually upgrade it. This is only the most recent
example of how traditions are being displaced before there is a full recognition
of their value to the quality of communal life. In communities where face-to-
face communication is still the norm, knowledge can be shared without having
to own a computer or pay a monthly fee to be "on-line." Unlike the abstract
knowledge of the new digitalized marketplace, knowledge acquired through
face-to-face communication is deeply contextual and requires a communal sense
of responsibility in how it is to be interpreted and used.
   There should be a place in the curriculum, especially at university level, for
examining another fundamental difference between modern, consumer-oriented
cultures and ecologically-centered cultures. This difference can be put into
perspective by considering the kind of ecological wisdom that is transgenera-
tionally communicated by elders and the kind of stories that older people tell in
modern cultures. After the basic differences are clarified for students, including
the deep cultural roots of the modern bias against all traditional forms of
knowledge (even though we continue to rely upon them in every area of daily
life), students should be encouraged to do a cultural inventory of the forms of
elder wisdom in their own community. This study of the local community and
place, which should bring into sharper focus the differences between older
people who tell stories of the American Dream of material success (or who apol-
ogize for what they now experience as the social irrelevance of their lives) and
elders who have learned to take on a unique responsibility for passing on
communal wisdom of human/nature relationships. This should be followed by
a discussion of the role of youth in the process of carrying forward and
renewing the ecological and cultural wisdom of previous generations. As I have
written elsewhere (Bowers 1995: 135–77), elders cannot contribute to the
process of community renewal if the younger generation does not know how to

participate in this very complex form of dialogue – or do not understand that they cannot be mentored in how to become an elder if they are using their time "surfing" the internet.

Conclusion

The globalization of modern culture has contributed to the spread of institutional values which threaten cultural and ecological diversity. As the communities in which students live largely incorporate these modern values and institutions, it is essential that the curriculum presents students the opportunity to develop a deeper understanding and appreciation of what Sim Van Der Ryn and Stuart Cowan refer to as "ecological design." That is, students need to acquire the vocabulary necessary for recognizing the essential relationships that are obscured by the industrial model of design. This vocabulary is largely contained in the following ecologically and culturally attuned design principles: (from Van Der Ryn and Cowen 1996: 54–5)

1   Design solutions must grow from an understanding of place.
2   Ecological accounting should be an integral part of the design process.
3   Design solutions should not violate the integrity of natural systems.
4   The knowledge of the community needs to be incorporated into the design process.
5   Design solutions should contribute to an increased awareness of the natural processes that sustain us physically and enhance us spiritually.

Not only does ecological design begin with local knowledge and traditions, it also includes the particularities of place – the climate, topography, soils, water, plants and animals, flows of energy and materials and other factors. The future of bioregional education is to articulate a design that can foster new relationships within a context and place, and to preserve the relevant ecological structure of the community.

There remain several challenges in the transition from a modern to a more bioregional form of education. The shift from basing educational reform on the different genres of liberalism to basing it on an orientation that balances cultural renewal with the need for ecological accountability will require radical changes in the education of public school teachers and university professors. A first major challenge will be to get the entrenched teachers and professors who control the form of knowledge (including the legitimating ideology and epistemology) that is to be learned at each level of the certification process to recognize the scale and accelerating nature of the ecological crisis. A second challenge will be to get them to recognize the culturally specific nature of the assumptions that underlie the high-status forms of knowledge under their control, and how these forms of knowledge have contributed to the colonizing of cultures that had previously taken a more ecologically oriented pathway. The third challenge will be to reorient their disciplines in ways that contribute to the ability of communities to develop more ecologically sustainable patterns of association and convivial activity, as well as technological practices. In light of the mounting evidence that the ever increasing flow of new technologies – and the accompanying mindset of techno-optimism – unleashed upon the world by the Industrial Revolution (along with supportive developments in the sciences, economics and political theory) are major factors contributing to the worldwide nature of the ecological crisis, the hubris that still characterizes all levels of the academic community does not provide a real basis for optimism. Bioregionalism provides a basis for recognizing how the more limited efforts to reform public schools and universities fit into the larger process of changing the course that modernization has put us on.

References

Bowers, C.A. (1993a) *Education, Cultural Myths and the Ecological Crisis: Toward Deep Changes*, Albany: State University of New York Press.
—— (1993b) *Critical Essays on Education, Modernity and the Recovery of the Ecological Imperative*, New York: Teachers College Press.
—— (1995) *Educating for an Ecologically Sustainable Culture: Rethinking Moral Education, Creativity, Intelligence and Other Modern Orthodoxies*, Albany: State University of New York Press.
—— (1997) *The Culture of Denial: Why the Environmental Movement Needs a Strategy for the Reform of Public Schools and Universities*, Albany: State University of New York Press.
Colborn, T., Dumanoski, D. and Myers, J.P. (1996) *Our Stolen Future: Are We Threatening Our Fertility, Intelligence and Survival?* New York: Dutton.
Gouldner, A. (1997) *The Future Intellectuals and the Rise of the New Class*, New York: Seabury.
Illich, I. (1971) *Deschooling Society*, New York: Harper & Row.
LaTouche, S. (1992) *In the Wake of the Affluent Society: An Exploration in Post-Development*, London: Zed.
Moravec, H. (1988) *Mind Children: The Future of Robots and Human Intelligence*, Cambridge: Harvard University Press.
Sachs, W. (ed.) (1992) *The Development Dictionary: A Guide to Knowledge as Power*, London: Zed.
—— (1993) *Global Ecology: A New Arena of Political Action*, London: Zed.
Shiva, V. (1993) *Monocultures of the Mind: Perspectives on Biodiversity and Biotechnology*, Penang: Third World Network.
Turkle, S. (1996) "Who Am We?" *Wired* 4.01: 148–52.
Van Der Ryn, S. and Cowan, S. (1996) *Ecological Design*, Washington D.C.: Island Press.

12 Bioregional restoration

Re-establishing an ecology of shared identity

*Michael Vincent McGinnis, Freeman House
and William Jordan III*

Sometimes the progress of man is so rapid that the desert reappears behind him.
The woods stoop to give him a passage, and spring up when he is past. . . . In
these abandoned fields and over these ruins of a day the primeval forest soon
scatters a fresh vegetation; the beasts resume the haunts which were once their
own; and Nature comes smiling to cover the traces of man with green branches
and flowers, which obliterate his ephemeral track.

(de Tocqueville 1991: 305)

Restoration and the mimesis of nature

How modern society has come to relate to place is the problem. Whether we
are talking about conserving, preserving or restoring nature, our treatment of
the natural world – expressed as unique, unexpendable, but interrelated places –
is essential. An allegiance to modern values and institutions continues as the
ideals of growth, development and technological progress spread from the West
to the East and from the North to the South. Major Western port cities, for
example, serve as the modern tributaries of a developing world's resources,
people and culture. To serve material and technological ends, the West exports
its bureaucratic institutions, ideals of growth, "environmental" values, beliefs,
technologies and machinery. Industrial capitalism has fostered the colonialization and homogenization of naturally and culturally defined spaces. The
resources, and sometimes the places themselves, are transformed into commodities for universal distribution. The question is how do we begin to restore a
sense of place?

In an extraordinary analysis of the importance of cultural mimesis, or reciprocal perception between species, Paul Shepard in *The Others: How Animals
Made Us Human* writes:

The declaration that "I am a fox" or that "you are a goose" is the predica-
tion of an animal on a pronoun which is more or less amorphous and helps
to teach the art of metaphor. Just as I say I may be foxy in strategy I can be
a tree in my rootedness or a rock in stolidity. Such multiple ritual assertions
are a kaleidoscope of successive, shared domains that define me ever more

precisely. *My identity is not simply human as opposed to animal. It is a series of nested categories.*

(1996a: 85; our emphasis)

There is no recognized "environment" in Shepard's characterization of an *ecology of shared identity*. Individual and cultural identities extended to include the lives of other species can serve to unite a culture with nature and place. This ecology of shared identity is a mirror of the multiplicity of place; a place that includes a circle of animals and habitats. In perhaps the most comprehensive inventory of how and why humanity relates to place, Shepard shows how a culture's ability to adapt and endure is dependent on context, modes of understanding and organizing, bioregional history, beliefs and values.

In globalization, our shared ecological identities are in jeopardy. As the bioregions which are our field of being are reduced by the extinction of species and the pollution of soil, air and water, humans suffer from a condition of diminished health and perception. As the well of natural provision dries up, we lose our inclination to treat each other generously. Our ability to analyze our situation and relationships with place is reduced. We find it difficult even to think clearly about our condition. Robert Harrison agrees and writes:

> Precisely at the moment when we have overcome the earth and become unearthly in our modes of dwelling, precisely when we are on the verge of becoming cyborgs, we insist on our kinship with the animate world. We suffer these days from a new form of collective anxiety: species loneliness.
>
> (1996: 428)

This is the condition within which the restorationist works: We are disabled creatures dislocated in a wounded landscape. How we organize to deal with ecological crises will determine our shared fate.

Species loneliness in a wounded landscape moves us to want to restore our relationship with place and others, or to put it another way, modern humanity yearns to re-establish and restore an ecology of shared identity. Rather than understanding the world through a relationship with earthly entities, modernity favors the human ability to experience nature as a quality (or quantity) that springs from scientific, technological, bureaucratic and economic understanding. Human beings remain isolated actors in an earthly cage: The world is technologically divided, scientifically categorized and manipulated, and is perceived absent of spiritual and intrinsic worth.

One value for restoration is that it provides a context of negotiating a relationship with nature and community. This chapter characterizes two paths of restoration – isolate and bioregional. Isolate restoration is based on the values of modern science which support the separation of place from culture, while bioregional restoration supports the reintegration of culture with place. Bioregional restoration requires active participation in the restoration of one's earthly home, and the relearning of lost social and community values. We propose that working to restore place is a way to preserve, enrich (and in some cases generate) a community's sense of place. When an inhabitory community engages the challenge of restoring its ecosystem functions, it is embarking on a self-educating, and culturally transformative path.

Two paths of ecological restoration

An act of restoration – be it the restoration of the Sistine Chapel, the reconstruction of a decaying downtown thoroughfare, or restoring the landscape to its "native" function – are mimetic acts if only because we are not quite sure what it is we are restoring. A "mime" is defined by *Webster's New Dictionary and Thesaurus* (1990) as "any dramatic representation consisting of action without words; a mimic or pantomimist.–*adjs* mimet'ic, apt to imitate; characterized by imitation.–*n* mimic, one who imitates, especially an actor skilled in mimicry.–*adj* imitative; mock or sham.–*vt* to imitate, especially in ridicule; to ape." Mimesis is a biological and a cultural phenomenon. Biologically, we find evidence of mimesis in the human immune system and the reproduction of DNA, antibodies and cells. The HIV virus (a nonliving entity) survives by masking itself as a healthy human cell; the virus imitates and copies the DNA patterns of a human being's cell structure to avoid antibodies. A culture's ability to mime is also an important function of a its ability to adapt.

If we are to mimic natural systems, what exactly are they? Little is known about the native speciation in the mosaic of ecological settings that preceded industrial exploitation. Some ecosystem types have disappeared altogether. For example, prairie plant and soil associations in the US have been so thoroughly altered for human purposes as to represent a nearly complete break in the historical continuity of ecosystem processes and functions. We do not suggest that cultural and natural histories are independent of one another. Rather, a place necessarily incorporates cultural and natural history or what Flores described in Chapter 3 as bioregional history. "Religion, art, ideas, institutions, and science through which a culture expresses itself are ultimately reflections of the ways it relates to nature" (Harrison 1996: 426). These various modes of social interaction change and vary, and are shaped by bioregional history.

Modern society's ability to mime nature no longer flows directly from daily observation. Rather, ecological restoration is a simulation of constricted reality. This is an important point because the restoration of "nature" or the "renaturalization of nature" is essentially a mediated activity. For example, modern society employs modern technology and science to restore some semblance of nature. So, in a scientifically oriented society like that currently prevailing in the West, we tend to think about an activity such as ecological restoration in scientific and technological terms. Scientific investigation, however, is conditioned by culture and context. It is important to keep in mind that, like science itself, restoration projects are carried out in the context of a discourse that is by and large conditioned by culture (McGinnis and Woolley 1997). Therefore, restoration ecology is essentially a mimetic simulation of an image of manipulated nature which includes some combination of nature and cultural artifice. In *Simulacra and Simulation* (1994), this type of activity is referred to by Jean Baudrillard as "simulacra" – ecological restoration represents a simulation of a simulation of an original nature. As Baudrillard writes: "We might believe that we exist in the original, but today this original has become an exceptional version of the happy few. Our reality doesn't exist anymore" (1995: 97). This is the same point that is made by Gertrude Stein when she writes: "A rose is a rose is a rose." Like nature, each form of the rose has changed meaning in time and context. The secrets of nature and the rose have been lost for good, *it seems*.

Ecological restoration is an active pursuit to understand and explore these "secrets" of nature which are hidden beneath the cultural screen of modernity. These secrets of nature can be exposed when we open ourselves up to the sensual and perceptual qualities of the living landscape. Modern society relies on scientific observation to expose the secrets of nature, not animism, aping or tribal activities. Modern science is based on a distinction between the self-mind and the object-nature. With these Greek didactic requirements, scientists propose that they can gain *objective* knowledge of things in nature. We place great faith in the individual (philosopher, scientist, consumer) to create and understand the nature of things. The primary medium of understanding is not the shaman, dance, song or ritual but rather modern science which is objectively oriented, in isolation from nature, politics and community. The scientific eye serves as the mediating force between the subject to be replicated and imitated, and the objective, material world. Myth-making and other "primitive" pursuits are looked down upon.

A clear example of our dependence on science as a mediating force between the subject and the object of discovery is found in the "scientific" aspects of exploration. The science of exploration is based on mapping. Modern maps have functioned as a mechanism for systematic exploitation of other lands and people. Charles Simpson argues:

> The mapping activity itself is a ritual of taking possession. It includes marking boundaries, recording and naming topographic features, and fragmenting a fluid landscape of human and animal vitality of life forces passing from nature to people and from generation to generation – into an abstract configuration of spatial coordinates and the domains of discrete sciences.
>
> (1992: 195–6)

Ecological restoration is also a scientific exploration of the past with the present in mind. As a process of exploration, the act of restoration should be vast and open, fluid and dynamic. Yet ecological restoration as a science often constructs the material world as a model-in-thought – an objectified experiment. We refer to this form of restoration as the *isolate* path. Generally, the isolate restorationist does not sense or feel an animate breathing world but, rather, accepts the prevailing mode of science which dictates a particular view or vision of nature treated as if it were an object to be observed and studied from the outside. Generally, the scientific bias toward breaking nature into parts to be studied in isolation has remained dominant in ecological restoration. Whether practiced as policies on public lands or mitigation efforts on private lands, human interaction with the dedicated places has been precisely defined and controlled either by a land-management bureaucracy or by legislation attempting to modify the excesses of industrial development. Economic practices and social biases surrounding the comparatively minuscule land areas devoted to laboratory models of ecological restoration tend to go on as before.

The National Research Council defines restoration as:

> the return of an ecosystem to a close approximation of its condition prior to disturbance. In restoration, ecological damage to the *resource* is *repaired*. Both the structure and function of the ecosystem are *recreated*. Merely recreating the form without the function, or the function in an artificial configuration bearing little resemblance to a *natural resource*, does not constitute restoration . . . *the goal is to emulate*.
>
> (NRC 1992: 1; our emphasis)

From this definition, it is clear that the human ability to mime and mime well is surely put to the test in restoration. How and why human beings restore nature is as important as what human beings restore.

The language used by the NRC can be interpreted as a definition of nature as a resource which can be repaired and recreated after it has been reduced or destroyed. While the focus on ecosystem structure and function represents a significant policy advance, it remains possible (and given the contemporary configurations of political process, likely) to confuse the NRC's definition of restoration with the legislatively driven industrial practices of *reclamation*. For example, because modern society values the salmon, human beings have technologically reproduced and augmented the declining numbers of salmon populations, fisheries and runs with over 100 hatcheries, which for most of their history have ignored the adaptive life cycles of wild stocks. Thus, hatcheries function as the machinery of subsidized reclamation – simulating natural abundance so that the public perception of the need for ecological restoration is clouded. Other examples of reclamation are familiar to anyone driving the highways of Europe and North America. Multinational corporations replace the dirt, soil and rock of a mountain that has been stripped of its ore in order to make the mountain aesthetically pleasing. A timber company replants a clear-cut forest with monocultures of commercially valuable trees. Each "reclaimed environment" is the embodiment of an image of how nature should appear after resource development. The question is, what has been restored – self-regulating wild systems or the *spectacle* of wildness?

These copies of nature are creations or, better yet, counterfeits, that, more often than not, support the future human use and exploitation of nature as a natural resource. The replacement of nature in the human image of the machine reinforces and perpetuates the instrumental values of re-use. The reforested landscape appears "prettier" than the "messy" old-growth forest ecosystem; the new forest is planted in rows which can be more effectively and efficiently clear-cut. Lost in the use and re-use of nature as a resource are the life-giving and self-organizing qualities of a healthy ecosystem. The replacement of nature by the best technologies and scientific information available falls short of the ideal of ecological restoration. Furthermore, the NRC's definition reinforces a view of wilderness as nature absent of human disturbance (with the possible exception of resource management experts) and, as such, it is a mirror image of the resource management biases out of which it has grown.

Ecological science and empirical observation cannot tell us what is natural. As Shrader-Frechette and McCoy show:

> One of the most common goals of ecologists is the attempt to specify and sustain what is natural . . . [but] ecologists cannot always specify what is "natural". . . . Knowing that one is acting in accord with nature is often defined as a condition in existence before the activities of humans who perturbed the system. . . . The definition is flawed, however, both because it excludes humans, a key part of nature, and because there are probably no fully natural environments or ecosystems anywhere. Because natural systems continually change, it is difficult to specify a situation at one particular time, rather than another time, as natural. [W]e are unable to define natural in a way *free of categorical values.*
>
> (Shrader-Frechette and McCoy 1994: 102–3; our emphasis)

Without intending to deny the enormous values of the science that drives ecological restoration, we must remain aware that restoration necessarily includes epistemic, cognitive and perceptual values. We propose that these values should be place-based.

The isolate restorationist finds it difficult to grapple with the continuum of human interdependence with a healthy, self-sustaining bioregion and community. Many individual scientists recognize that the health of human communities and the health of bioregions are coterminous, but the tools of their trade – their reductionist rigor – often prevent them from engaging the messy interpenetration of humans and places without violating their own discipline. Rather than engaging the interpenetration of human communities and the landscapes they inhabit, the scientific "eye" embedded in the body of isolate restoration efforts can serve to reinforce the subject/object separation of place and human community. Donald Worster, in his definitive study of the development of ecological thinking, *Nature's Economy* (1979), illuminates this trend in the discipline of academic ecology in the latter half of the twentieth century:

> But at the very moment [ca. 1949] he [Aldo Leopold] embraced it as the way out of the narrow economic attitude toward nature, ecology was moving in the other direction, toward its own niche in the modern tech-

nological society. It was preparing to turn abstract, mathematical, and reductive.

(1979: 289)

Despite our technological intentions, place continues to influence human activities and cultures in countless ways – altering our habits, cities, cuisine, language, values and expectations. Moreover, the loss of a species or a mountain is not merely a failed experiment. The loss of a species represents the diminishment of our perceptual field of vision.

We should attempt to restore the human relationships and shared perceptions that define a community of place. A mimetic relationship that blurs the boundaries between the subject and the object, and the division between place and society is needed. In such a relationship, "There will no longer be a humanity, or a nature, but a continuum of connection that is the primal asking force" (Rothenberg 1996: 265).

We propose a second path for ecological restoration – *bioregional* restoration. The goal of bioregional restoration is to reimmerse the practices of human community within the bioregions that provide their material support, as well as the direct relationships to the more-than-human world on which the full range of human experience depends. Bioregional restoration is a performative, community-based activity based on social learning and cooperation. If inhabitory communities are left out of the process of restoring the landscape and place, then restoration is not bioregional. Bioregional restoration can be a therapeutic strategy to expose ourselves viscerally to local ecosystem processes, to foster identification with other life-forms and to *rebuild* community within place, as the insights and local information that emerge from restoration activities affects the cultural and economic practices of the human population.

Some differences between the values of isolate and bioregional restoration are described in Table 12.1.

Note that the distinctions between the isolate restoration path and the bioregional restoration path are not mutually exclusive. If the discipline of

Table 12.1 A comparison of restoration paths

	Bioregional restoration	*Isolate restoration*
Function	Communion	Observation
Relational	Community practice	Management based on expertise
	High degree of interpenetration between cultures and places	"Nature" studied in isolation from human influence
Technology	Focus on locally appropriate technology	Coexistence with industrial production
Science	Experimental	Data-based
Activity	Preservation	Replacement
	Cooperation	Domination

Source: McGinnis 1996

restoration ecology had not risen independently, attainment to reinhabitation would have required its invention. Any restoration effort that attempts to engage the industrialized landscape is dependent on the insights of ecological science. The practitioners of bioregional restoration, grounded in particular places, rarely have the luxury of isolating the human presence from the more-than-human world; neither can they afford to eschew the powerful tools that science provides. Thus we will find the value differences described in Table 12.1 converging.

The goal of promoting shared living place is not the goal of isolate restoration. Scientific evidence supports the reintroduction of the wolf to the Greater Yellowstone Ecosystem. The wolf is viewed as an important predator in the Yellowstone greater ecosystem. Restoration of the wolf to Yellowstone, however, requires more than the reintroduction to the wolf to the system. The sensual, sacramental, spiritual, and ecological values that are endemic to a healthy wolf population and the wolf's place in a diverse Yellowstone bioregion should be respected, cared for and ultimately restored. Relationship-building is not a primary concern for the isolate restorationist. At the same time we wish to restore the wolf to Yellowstone, we also support a public policy that led to the murder of nearly one-third of the total bison herd that trespassed the boundaries of Yellowstone Park in search of food. In contrast to isolate ecological restoration, bioregional restoration requires an alternative relationship with these Others.

The bioregionalist begins from a different set of motivations and constraints than does the isolate ecologist. Place may be scientifically defined by its geomorphic, ecological and hydrological characteristics. For the bioregionalist, the scope is expanded to include the degree to which local communities enfold themselves within the constraints and opportunities of particular places. Human cultural definition from within the bioregion plays as large a role in the definition of place as do the more quantifiable nonhuman aspects identified by isolate science (Zuckerman 1992). Bioregional restoration is a *practice* performed by a community that extends its identity to biospheric life as manifested by particular places; a human community which begins to define itself through its continuity with and immersion in ecological systems.

Bioregional restoration and community-building[1]

The key to understanding bioregional restoration lies in the recovery and reconstitution of the human community. What is community? How is community created? What might the act of ecological restoration have to contribute to building and sustaining community?

As it happens, the word "community" provides a convenient way of approaching these questions. At first glance, one might suppose that this word is derived from the Latin word "unus," or "one," and that, together with the prefix "cum" (with), it means something like "all together in union" or "at one with." This, however, is not the case, and the point turns out to be a crucial one with important implications for understanding the value of bioregional restoration.

In fact, the word "community" derives not from the word for "one," but from another Latin word, "munus," which has an extremely interesting range of meanings, including service or duty; gift; and sacrifice. The word "community," in other words, is a metaphor. At its root is the idea of an exchange of services – out of duty, it may be, but also, pointing to another dimension of the idea, freely, even affectionately, as a gift, or even a sacrifice. A community, then, is the assemblage of individuals to whom one is bound by this kind of relationship – one defined, we might even say constituted, by mutual obligation and by an exchange of gifts.

What, then, does this have to do with the act of ecological restoration? At least part of the answer to this question is obvious. Bioregional restoration is, first and foremost, a service we offer to nature and to each other. And at the same time, by giving us work to do in the landscape, it satisfies the first requirement of membership in the land community. We will return to this notion of working with the landscape later.

Bioregional restoration is not only a service we offer to nature out of a sense of duty, it is also in many instances a gift, offered freely out of love and affection for a place. Gifts, as the second meaning of the root-word "munus" suggests, play a key role in establishing and defining a relationship or rebuilding a community, but, as the poet and philosopher Frederick Turner has pointed out in his book *Beauty: The Value of Values* (1991), an exchange of gifts is always to some extent problematic because we can never be sure that what we give is commensurate in value with what we take or have been given. This is true in the case of our relationship with the rest of nature: since nature gives us all we have, including life itself, how can we ever pay it back in kind?

Surely it is just this misgiving, this uncertainty about the value of the restored ecosystem, its authenticity and its worthiness, that underlies the ambivalence that many have about the promise of restoration. Presuming to give something back to nature, to repay nature in kind for what we have taken from it, restorationists find themselves in the classic position of givers of gifts: the painful condition of never being sure that what they give is good enough or worthy of its recipient.

Faced with this prospect, the isolate restorationist often responds by avoiding the exchange of gifts. The view of the landscape as a laboratory to be left alone may be interpreted as an attempt to avoid the troubling predicament of the gift-giver. But, since we cannot avoid taking from nature, to refuse to offer any gift in return is to define ourselves purely as consumers of nature – parasites on it.

Even worse, to refuse to offer a gift back to nature is to cut ourselves off from communion with it. Bioregional restoration represents a culture's attempt to give back to nature in kind in return for what we have taken from it, and if we find that troubling, it is important to remember that the exchange of gifts – the very act that defines a relationship – is always and necessarily an uncertain act in which we encounter the limits of our own abilities and the ambiguity of the relationships we seek to restore.

We tend to think that relationships and communion are easy, natural and perhaps free of tension and uncertainty. This view, however, is clearly a peculiarity of our own modern civilization. Premodern cultures continue to view community and nature differently – not as easy, but as perilous and uncertain, and not as "natural" but as an achievement of the human community acting together to confront this uncertainty and find ways of coping with it.

This is clearly evident from the ritual technologies premodern cultures have developed to perform the work of community-making, and, as described earlier, is carefully explored by Shepard (1996a) in his portrayal of the cultural mimesis of the natural world. Rituals of mimesis and initiation, by which a child achieves membership in the human community, often involve humiliation, ritual death and the mutilation of the body. The rite of communion itself, by which the human community negotiates its relationship with the more-than-human world, begins in the act of killing, and represents an attempt to come to terms with the fact that life depends utterly on death. Destruction of "nature" is very much a part of restoration activity – as exemplified by the removal of "exotic" species from a riparian area. Hence, destruction and construction are part of restoration work.

It is this observation that reminds us of the significance of the third meaning of "munus" – sacrifice. Contained in the word is the realization that community – or communion – depend not only on sympathy and the more congenial aspects of relationships, but also on the negative ones as well – the crisis of identity, the violence against the Other, the contamination of the self that is inseparable from real communion. It is for this reason that it is the act of killing and eating that provides the idea of communion. This is the wisdom of human culture – that in its attempt to negotiate relationships with others, it goes directly to the most obviously problematic kind of relationship – that between predator and prey – and to the crisis inherent in it.

What we propose is that rebuilding community is a perilous, difficult endeavor, in which we encounter tensions that cannot be avoided, but that are inherent in the experience of community. Plans to remove "non-native" plants or animals with the hope of returning native inhabitants to the landscape are widespread, and necessarily involve an exchange of values. On Angel Island in the San Francisco Bay, commercial loggers have been busy chopping down some sixty-four acres of (more than 12,000) eucalyptus trees, which are non-native to the island. Biologists say that these introduced trees have been crowding out native plants, and there is evidence that the native grasses, wild flowers and shrubs are returning. Restoration, in this case, represents the exchange of the value for the native wild landscape that exists underneath the dominating but aesthetically valuable eucalyptus forest. These are the tensions that make us feel ambivalent about the act of restoration.

Encountering these tensions, we tend to deny them: This is not what we expected a healthy relationship with nature to feel like. But what we have to

remind ourselves at such moments is that there is every reason to expect relationships and community to feel like that – that indeed the great value of
bioregional restoration is precisely that it draws us into this uncomfortable
confrontation with our limitations and conceptual boundaries, and like the
rituals of initiation and communion, provides a context in which to come to
terms with them.

Hence the key role of bioregional restoration is the building of a human
community, the self-definition of which is extended to include the larger biotic
community. Place-based ecological restoration can provide the shared experience, knowledge and ritual necessary to such an undertaking. This is not a
solitary experience, but rather lends itself to group effort and even toward celebration and festival. Bioregional restoration must not only deal with the
historical degradation of ecological processes due to human practices but with
the artificial boundaries that separate the inhabitant from his or her own local
habitat, and with the variety of values represented by human residents.

However much our thinking minds may spin off in the direction of technological invention and the comforts of a controlled environment, our senses
remain immersed in a bioregion which is not entirely of our construction or
invention. The bioregion is the source of our deepest pleasures. Our current
alienation from these processes may not be as profound as we sometimes
fear. The initiation of community-based ecological restoration projects is a
powerful context in which to put the combined tools of science and bioregional sensibility to work at the service of personal and community
transformation – to begin the process of reorientation of inhabitant to
habitat, of community to place. Such projects inevitably lead their practitioners to an ever-deepening collective experience of the processes of place,
and move them to confront the barriers that separate them from it. Keith
Basso points out, "place-making is . . . a way of constructing social traditions
and, in the process, personal and social identities. We *are*, in a sense, the
place-worlds we imagine" (1996: 1). Bioregional restoration offers the opportunity to imagine ourselves back into our place-worlds while maintaining the
evolutionary continuity of the communities of flora and fauna which define
the particularities of each place.

The value of shared service

The scale at which the bioregional activist defines a project is logistically important. The most decisive factor may not be abstract descriptions of
biogeographical provinces but the perceptual limitations of humans living
within them. The watershed lends itself well to these limitations of human
perception; it is a visible hydrological container of all our coexistent life-forms;
it is what lies between our eyes and the horizon. If the watershed or the
ecosystem is an essential unit of cultural perception, we must turn our attention
to the body of perception, the spider's web of language, ritual and vernacular
life which constitute the reciprocal reinforcement of individual and collective

identification with the processes of place. As if to accommodate the varieties of human skills and energies, the watershed breaks itself into ever-smaller increments – river to creek to swale. A prerequisite for bioregional living is the development of a healthy relationship between humanity and these various parts of a watershed. The proliferation of inhabitory watershed and creek councils over the last twenty years illustrates the efficacy of the watershed as a unit of perception, both in terms of the restoration of ecosystems and the revival of functional human community.

In the Pacific Northwest, watershed councils have more often than not had their genesis in the desire to restore local runs of wild salmon. The Pacific salmon in this ecoregion has recovered much of the totemic significance it had for first peoples (House 1974). For modern communities, the discovery that individual salmon stocks are exquisitely adapted to the constraints and opportunities of particular watersheds often serves to illuminate the need for new-old models of human self-organization. Direct engagement with the survival of specific races of salmon opens the doorway to our primary institution of higher learning: the natural world as articulated by locale. The systematic examination of the root causes of habitat degradation required by any honest effort at the rehabilitation of salmon stocks leads inexorably to an examination of the human economic practices that lie behind them; dams in one watershed, rapid deforestation in another, water diversion for agriculture in yet another.

To re-experience John Muir's revelation that every part of nature is connected to every other part does not require the presence of a charismatic species such as salmon. Communities which engage any single part of the self-healing attributes of bioregions are inevitably drawn into the multitudinous relationships that define life in place. The decades-long discovery and restoration of the prairie, savannah and forest ecosystems practiced by the North Branch Project of suburban Chicago has lead to the elevation of the roles of fire and the presence of buffalo in ecosystem function as elements of political and social discourse in the North American Midwest (Stevens 1996). Approaching the same ecoregion from another vector, the Land Institute's work in developing a sustainable agriculture based in prairie ecosystem restraints has increasingly come to be seen as an essential element in the survival of human community there (Jackson 1994). The success of the Monday Creek Restoration Project, one of the first attempts to reclaim the one-third of Appalachian streams rendered lifeless by acid-mine drainage, may depend as much on the fact that it is embedded in the community revival efforts of Rural Action – a community service organization – as it does on capturing a large federal grant to undertake its daunting task. The restoration of autonomous, self-regulating, self-reliant communities is inseparable from the restoration of ecological systems. Bioregional restoration efforts, when successful, will inevitably lead their practitioners away from the popular political abstractions that have come to describe the tensions between an economic or ecological interpretation of the natural world (i.e. jobs versus environment) into shared strategies for the re-creation of place-based, enduring community.

Any bioregional restoration project quickly discovers its dependence on maps. Land-use histories must be understood at the scale of the landscape; the degradation of habitats needs to be traced to its sources; relatively undisturbed areas need to be located and assessed for their potential as refugia. If the watershed is the chosen context, a rough understanding of hydrological and geomorphological processes of entire drainages must be assessed in order to ascertain the most effective point of application for rehabilitation projects. Until the early 1990s, few of the geophysical maps available to the restorationist provided information – essential to the work at hand – about the life-processes of places. Undertaking the effort of reorganizing available cartographic data into the context of natural areas is a valuable exercise. It quickly reveals local information gaps, the filling of which becomes an early priority for the restorationist. When the informational gaps are filled by residents trained in systematic observation, several transformational processes are set in motion.

The first of these is that inhabitory people will find themselves participating in Thomas Berry's third principle of bioregional function (1988), that of *self-education*. Inhabitory mappers bring to the process of mapping a vernacular familiarity with the processes they are observing which transforms lifeless data into a living extension of the experience of place (Aberley 1993). As these experiences are collectivized in the form of maps, cartography is invested with new functions and powers: a tool developed as a rigid expression of property becomes an expression of life processes. Maps now become a tool for turning sensually and topographically limited individual perception into communal place-knowledge; locally drawn maps become the externalized mental grid within which the fluid conditions of the more-than-human community can be considered. Maps oriented to ecologically or hydrologically discrete areas, constructed from *within* the area, create a new context for social discourse that has previously been obscured by the procedural organization of governments and/or the pressures of the marketplace.

As locally developed maps are revisited over time to monitor changing conditions resulting from either economic activities or restoration projects, a temporal dimension is added to perception of place. Through this process, the bioregional community becomes entrained in the dynamics of natural succession, and learns more about the effulgence of natural healing processes. The human community begins to experience itself *in time*, to understand itself as part of a *process* of which it is an instrumental part. Such maps contribute to the logic of new social institutions based on a shared ecological identity.

The Mattole Restoration Council (MRC), a bioregional restoration project based in the Mattole River watershed in Northern California (House 1990; 1992), had its beginnings in a community-based effort to maintain and restore one of the last native chinook salmon stocks remaining in California. Engagement in the enhancement of spawning populations has taught its practitioners that there is a direct and reciprocal relationship between the processes of mature forests and the health of aquatic ecosystems. Despite the fact that forest practice rules that require an assessment of the cumulative effects of logging had been in place in California for more than a decade, no regulatory agency (or anyone else) had ever mapped the degree to which the mature forests in the watershed had been reduced. Using historical aerial photographs and county tax records, the MRC produced a poster-map demonstrating that in a mere forty-year period preceding 1988, 93 percent of the ancient Douglas-fir forests had been extracted with little attention paid to ecological consequences to the watershed. The poster map was mailed to every resident and landowner in the valley with a text that proposed no strategy beyond a mild suggestion that a moratorium be put in practice on the logging of old-growth forests until forest land-managers could demonstrate a methodology that retained the basic elements of the ecological functions of late seral forests. At the time the map was distributed, very little of the remaining old-growth forest had any legal protection. Ten years later, two-thirds of the 7 percent of the old forests had been given the status of refugia in perpetuity through a combination of public discourse on public lands and aggressive land trusts established by local inhabitory forest constituencies to purchase private lands for public management. The remaining third of the ancient forest remains a subject of heated debate. Knowledgeable observers of the restoration efforts in the Mattole credit these refugia with being among the most important achievements of a two-decade effort which has included steady work by the resident population in the enhancement of salmon populations and habitat, erosion control, education and reforestation.

Bioregional restoration in the Mattole River watershed is being carried on at a more or less matter-of-fact level – the level of shared service. Such work produces teams rather than communities, but it is clearly a crucial step toward the creation of community. Building a sense of community and place necessarily involves the self-conscious exploration of the act of restoration as an exchange of gifts. The social tensions and ecological ambiguities this exchange entails become embedded in ongoing community discourse as inhabitants begin to understand themselves as instrumental parts of watershed function. As landscape rehabilitation work becomes a significant aspect of the local economy, the lessons learned from that work become incorporated in economic land-use practices that can be described as restorative. The philosophical goals of the MRC are based on communion with the watershed; the incremental growth of a collective understanding of what it means to be part of a system of human and ecological relationships.

The long path home: Working to cooperate with the land

> to leave the human body
> to the light of nature
> to plunge it alive into the gleam of nature
> where the sun will wed it at last

> (Artaud 1965)

Community engagement in bioregional restoration teaches the community some of the things it needs to know as it seeks to rediscover its adaptive relationship to place. Restoration does not lie in our rational interpretations of computer data, satellite images, or directives from the Capitol, but rather in our organismic immersion in the systems that operate within us and surround us. Ecological restoration, when practiced in the context of local integration of human communities with the functions of naturally unfolding regions, offers hope of a sustainable existence. As Jim Dodge notes: "Restoration, like any art, seeks a greater understanding of existence, which tends to deepen our appreciation, gratitude, and humility, salubrious states of mind that are less fringe benefits than compelling requisites for further work" (1991).

We can find these working relationships if we begin to understand that we are part of *autopoietic systems* (Margulis and Sagan 1986; Kauffman 1995; McGinnis 1996; Hayles 1996). An autopoietic or self-organizing system is limited by its boundary insofar as its structures and processes – its very existence – are based on its self-producing capacity. A bureaucratically constructed boundary fragments and divides a river into parts for "environmental management" (e.g. the fish and wildlife are separated from the mountain landscape which is separated from the riverine ecosystem). A boundary of an autopoietic system, like a cell, is open and emerges as the system's components interact. In an ecological system, the *unity* among the system's parts and its *wholeness* are called autopoietic (see the discussion of *autopoiēsis* in Chapter 4 by McGinnis).

The work of Shepard (1996a; 1996b), Thomas Berry (1988) and a very few others suggests that the multitude of human cultural expressions may have evolved autopoietically as an adaptive response to place. Advocates of an ecological anthropology such as Roy Rappaport have generated controversy and have been roundly attacked as "environmental determinists." While modern technology has allowed us to examine the very basic elements of physical evolution, we have yet to develop the tools that allow us to understand the incremental developments of cultural evolution. As an example, consider the Klamath River watershed of northern California. Tribal peoples in the Klamath basin arrived separately in the area over long intervals of time, establishing territories along the length of the river. The three tribes who lived there (and live there still) at the time that anthropologist Alfred Kroeber arrived not only spoke different languages, but languages of entirely different groupings. Yet by the time of Kroeber's arrival, common cultural practices had evolved centering around the river and its natural provision. The Yurok, the Karok and the Hupa peoples had evolved an elaborate and coordinated practice of ritual self-regulation in regard to the great salmon runs that provided (and provide) a large part of their sustenance. Similar rituals, timed according to the salmon's arrival at tribal territories, not only recognized and protected the salmon's reproductive requirements, but assured that fish would reach the peoples upstream (Kroeber 1925; Waterman and Kroeber 1938). Historian Arthur McEvoy notes that "their complex economic strategies did not emerge and did not endure simply as a matter of chance. They developed it, over time and no doubt at some cost, and maintained it deliberately" (1986: 10). Speculation regarding the nature of the negotiation and evolution of common and mutually beneficial rituals over a time-span of generations, among peoples speaking different languages, yields obvious parallels to the ambitions of the contemporary bioregional restorationist. Self-conscious social change in the direction of ecological adaptation and negotiated ritual, while more complex under the conditions of modernity, will no doubt be as much a determinant in the endurance of contemporary cultures as it was (and is) for the first peoples of the Klamath.

Ethnographers offer us some clues as to the enormously complex relationships of successfully adapted indigenous cultures, relationships intrinsic in language and the vernacular of daily practice and ritual, the specifics of which can be as nearly invisible as are the mycorrhizal relationships between plants and soil in a healthy forest. The clues that would lead us to understand the disappearance of maladaptive cultures are lost to us in much the same way that a desertified ecosystem gives us little indication of its former "self." Richard Nelson describes his "native natural history," *Make Prayers to the Raven* (1983), as a "guidebook to the boreal forest," to indicate the degree to which he discovers the seamless integration of people and place in his study of the Kuyokon people of central Alaska. "Apache constructions of place reach deeply into other cultural spheres, including the conceptions of wisdom, notions of morality, politeness and tact in forms of spoken discourse, and certain ways of imagining and interpreting the Apache tribal past," according to ethnographer Basso (1996: 15) in his account of the Cibecue region of Arizona. Basso quotes Cibecue elder Annie Peaches: "The land is always stalking people. The land makes people live right. The land looks after us. The land looks after people."

Bioregional restoration and cultural/ecological renewal are part of the same autopoietic process. Because human beings are part of this process of renewal and reproduction, restoration should break from a view of the moral, scientific and intellectual authority of the autonomous and isolated self. As members of communities, human beings should build a deep and lasting relationship with the natural, life-support system. Hard work should be recognized as a necessary part of the restoration process. Bioregional restoration means coming to terms with the meaning of work and one's labor in the community. As Richard White writes: "We cannot come to terms with nature without coming to terms with our own work, our own bodies, our own bodily labor" (1995: 171). In the context of bioregional restoration, the meaning of work begins to transcend a merely economic definition; a fuller, richer range of perceptual and communal opportunity is unleashed. The work of humans is contextualized by the work of rivers and watersheds, as they maintain their dynamic equilibrium in geologic time; by the work of forests and grasslands and deserts; by the intricate balance of cooperation and competition that defines the work of all species in evolutionary time. To paraphrase Peter Berg (1998, in progress), we are invited to go beyond making a living – to living a making.

In place, we can learn from the land and our coinhabitants there, educate ourselves and work with others to rebuild new-old relationships embedded in the gifts of the commons.

Acknowledgments

McGinnis thanks the National Science Foundation for financial support to investigate "the place of values and science in restoration." A version of this paper was presented at the Fourth Annual University of Florida College of Law Public Interest Environmental Conference in March 1998.

Notes

1 This section was written by Jordan. The remaining sections of this chapter were written by House and McGinnis.

Bibliography

Aberley, D. (1993) *Boundaries of Home: Mapping for Local Empowerment*, Gabriola Island: New Society Publishers.

Artaud, A. (1965) "Seven Poems," *Artaud Anthology*, San Francisco: City Lights Books.

Basso, K.H. (1996) *Wisdom Sits in Places*, Albuquerque: University of New Mexico Press.

Baudrillard, J. (1994) *Simulacra and Simulation*, trans. S.F. Glaser, Ann Arbor: University of Michigan Press.

—— (1995) "The Virtual Illusion: Or the Automatic Writing of the World," *Theory, Culture & Society* 12: 97–107.

Berg, P. (1998) *Creating a Bioregional Identity: The Selected Essays of Peter Berg*, work in progress.

Berry, T. (1988) *The Dream of the Earth*, San Francisco: Sierra Club.

de Tocqueville, A. (1991) *Democracy in America*, ed. P. Bradley, New York: Vintage Books.

Dodge, J. (1991) "Life Work," in *Helping Nature Heal: An Introduction to Environmental Restoration*, Berkeley: Ten Speed Press.

Harrison, R.P. (1996) "Toward a Philosophy of Nature," in W. Cronon (ed.) *Uncommon Ground: Rethinking the Human Place in Nature*, New York: W.W. Norton.

Hayles, N.K. (1996) "Simulated Nature and Natural Simulations: Rethinking the Relation between the Beholder and the World," in W. Cronon (ed.) *Uncommon Ground: Rethinking the Human Place in Nature*, New York: W.W. Norton.

House, F. (1974) "Totem Salmon," in V. Andruss *et al.* (eds.) *Home! A Bioregional Reader*, Philadelphia: New Society Publishers.

—— (1990) "To Learn the Things We Need to Know," in V. Andruss *et al.* (eds.) *Home! A Bioregional Reader*, Philadelphia: New Society Publishers.

—— (1992) "Dreaming Indigenous," in *Restoration and Management Notes* 10: 1.

—— (1993) "Watersheds as Unclaimed Territories," in D. Aberley (ed.) *Boundaries of Home: Mapping for Local Empowerment*, Gabriola Island: New Society Publishers.

—— (1996) "Restoring Relations," *Restoration and Management Notes* 14: 1.

Jackson, W. (1994) *Becoming Native to this Place*, Lexington: University Press of Kentucky.

Kauffman, S.A. (1995) *At Home in the Universe: The Search for the Laws of Self-Organization and Complexity*, New York: Oxford University Press.

Kroeber, A.L. (1925; 1976) *Handbook of the Indians of California*, New York: Dover.

Margulis, L. and Sagan, D. (1986) *Microcosmos: Four Billion Years of Microbial Evolution*, New York: Summit Books.

McEvoy, A.F. (1986) *The Fisherman's Problem: Ecology and Law in the California Fisheries, 1850–1980*, Cambridge: Cambridge University Press.

McGinnis, M.V. (1996) "Deep Ecology and the Foundations of Restoration," *Inquiry* 39: 203–17.

McGinnis, M.V. and Woolley, J.T. (1997) "The Discourses of Restoration," *Restoration and Management Notes* 15 (1): 74–7.

National Research Council (1992) *Restoration of Aquatic Ecosystems: Science, Technology and Public Policy*, Washington, D.C.: New Academy Press.

Nelson, R.K. (1983) *Make Prayers to the Raven*, Chicago: University of Chicago Press.

Rothenberg, D. (1996) "No World but In Things: The Poetry of Naess's Concrete Contents," *Inquiry* 39 (2): 255–72.

Shepard, P. (1996a) *The Others: How Animals Made Us Human*, Washingon, D.C.: Island Press.

—— (1996b) *Traces of an Omnivore*, Washington, D.C.: Island Press.

Shrader-Frechette, K.S. and McCoy, E.D. (1994) *Method in Ecology: Strategies for Conservation*, Cambridge: Cambridge University Press.

Simpson, C.R. (1992) "Mapping an Extreme Landscape," in J. Kleis and B. Butterfield (eds.) *Re-Naming the Landscape*, New York: Peter Land.

Stevens, W.K. (1996) *Miracle Under the Oaks: The Revival of Nature in America*, New York: Pocket Books.

Turner, F. (1991) *Beauty: The Value of Values*, Charlottesville: University Press of Virginia.

Waterman, T.T. and Kroeber, A.L. (1938) "The Kepel Fish Dam." *University of California Publications in American Archaeology* 35: 6.

Webster's New Dictionary and Thesaurus (1990) Concise Edition.

White, R. (1995) *The Organic Machines*, New York: Hill & Wang.

Worster, D. (1979) *Nature's Economy*, Garden City: Anchor Press/Doubleday.

Zuckerman, S. (1992) "Four Reasons Why You've Never Seen a Map of the Northern California Bioregion," in D. Aberley (ed.) *Boundaries of Home: Mapping for Local Empowerment*, Gabriola Island: New Society Publishers, 57–9.

Index

Bowers, Chet A. 7, 8, 191–204
Braudel, Fernand 51
Bryan, Frank 91
Buddhism 124
bureaucracy 64–6, 70, 73
Butzer, Karl 49

Caldwell, Lynton 71
California 87–8, 110–11, 142–3, 145, 217–18, 219
Callenbach, Ernest 24
Canada 5, 34, 141, 166
capitalism 36, 49, 64, 205
Carpenter, J. 28
Carr, M. 31
Carroll, Bob 24
CCHW *see* Citizens Clearinghouse for Hazardous Wastes
Central America *see* Mesoamerica
Cézanne, Paul 76
Chaliand, G. 123
Champlain-Adirondack Biosphere Reserve 86–7
Chicago 216
Chisholm, Donald 113
Chole Maya 175
Christianity 33, 64
Chumash tribe 1
cities 26, 198
Citizens Clearinghouse for Hazardous Wastes (CCHW) 164
civil society 101
class 192
climate: bioregional history 51; global warming 6, 102, 103, 133–50
co-operation, informal governance systems 113–14
cognitive land-use mapping 180
cognitive maps 75
collective action 106, 107, 111, 137
colonies 90, 176
commoditization 192, 195, 197, 200, 201, 202, 205
communications, globalization 6
"communicative action" 7
communitarianism 158, 174
community: bioregional restoration 2, 7; democracy 91; diaspora 122–4; functions 32; identity 8; knowledge sharing 165–6; low-status knowledge 198; NIMBY factor 163–4; place relationship 114; resource management 181–2; resource regimes 103–4, 105;

restoration 206–7, 211, 212–15, 216, 217–18, 219; transgenerational knowledge 202–3
complexity theory xvi–xvii
conservation: "conservation conundrum" 171–4, 185; local groups 87, 105, 110–11, 139–40, 164, 166–7; Mesoamerica 175–85; *see also* reserves
conservatism, cultural bioconservatism 191, 198, 199–200, 201
constructivism 110, 157, 159, 160
consumerism 198, 202; *see also* commoditization
consumption 94
control, bureaucratic 64–6
cosmopolitan bioregionalism 6, 8, 121–32
counterculture 14–15
Cowan, Stuart 199, 203
Cowdrey, Albert 53
Coyote, Peter 20
Cronon, William 49, 53, 83
cultural adaptation theory 49
cultural assumptions 193, 194–7, 198, 200–1, 203
cultural-conservatism 191, 198, 199–200, 201
culture: bioconservatism 199–200; bioregion-based 37; boundaries 17, 49–50; constructivism 157, 159; education 200–3; indigenous peoples 15–16, 161–2; literacy 196–7; low-status knowledge 199; mechanization 66; nature relationship 1–2, 21, 48–9, 51–2, 162, 206; place relationship 38, 46; restoration 207–8; shared rituals 219–20; technological transition 193–5
customs, adaptation to ecosystems 26
cyborgs 62–3, 64
Czempiel, Ernst-Otto 102

dams 63
dance 1, 24, 26
Darien Gap 175, 178
Dasmann, Raymond 13, 22–4, 163
Davis, Mike 66
de Tocqueville, A. 205
deBuys, William 54
decentralization 13, 29, 30, 36–7, 113
decision-making 133, 137, 147
deep ecology 33
deep time 4, 51, 54
deforestation 134, 148, 176, 181, 214, 218

degradation 74
democracy 91
Descartes, René 194
design, ecological 199, 203
determinism, environmental 43, 48, 162
devolution xv
dialogue, place-based scientific knowledge
    165–6
diasporas 121, 122–4
"Diggers" 18
Dodge, Jim 24–5, 29, 219
Dolcini, M. 27
Doughty, Robin 54
Douglas, Mary 161–2
Dryzek, John 72
Dust Bowl 49, 50, 52, 53

earth spirituality 33
Eco, Umberto 63
ecofeminism 33–4
ecological footprints 36
"ecological revolutions" 52
The Ecologist 22
ecology: bioregional delineations 47–8;
    movement development 15; restoration
    210–12; self-organization 71–2, 74;
    systems theory 49
economics: bioregional world view 37;
    bureaucracy 65; ecological decline 64;
    globalization 2, 67–8; green 35;
    industrialization 3; multinationals 67
ecoregions, boundaries 46–7, 49–50,
    88–9, 93
ecosystems: *autopoiesis* 219; mimesis 207;
    reclamation 209–10; systems theory
    49; technological modelling of 68;
    watershed boundaries 86
education 7, 191–204; environmental 87,
    95; self 32, 218
El Pilar 182–4, 185
Elder, John 95
Emberra people 175, 178
emissions, greenhouse gases 103, 134,
    135, 139, 140, 141, 143
energy use 134, 138, 140, 142–3, 144–5,
    148
environmental crises 16, 36, 102
ethnography 220
Europe 5, 35
evolution, adaptation 49, 51, 52, 61
exterior, interior relationship 75–6
extinction 82, 83, 122

Falk, L.L. 84
Feldman, David L. 5, 6, 8, 133–54
feminism *see* ecofeminism
"figures of regulation" 26
Fike, Michelle Summer 33–4
FitzSimmons, M. 5
Fleck, Ludwig 159
Flores, Dan 4, 8, 43–58, 207
folklore 2
Foltz, B.V. 72
forests: Maya 175–6, 181, 182, 185;
    Mesoamerica 173; rivers relationship
    217–18; Vermont 82–3, 84–5; *see also*
    deforestation; reforestation
Forsey, H. 36
Fox, Warwick 72
France 141, 144
Fraser, J.T. 126
"Free City" movement 19
"free rider" problem 137
Frenkel, S. 31
Friedmann, J. 30
Friends of the Mad River 87
Frisco Bay Mussel Group (FBMG) 20
frontiers 44
Fukuoka, Masanobu 198
functionalism, local authorities 108
"future primitive" 21, 37

Gaia concept 29, 130
Garifuna people 175, 176, 180
GCMs *see* General Circulation Models
Geddes, Patrick 30
General Circulation Models (GCMs)
    136–7
geographic information systems (GIS) 88,
    166, 180
geology, natural borders 88, 93
Gergen, David 125
Germany 141
Gibbs, Lois 164
GIS *see* geographic information systems
global civil society 6, 8, 101–2, 110–14,
    115, 129
global economy 122, 124–5, 126, 129
Global Rivers Environmental Education
    Network (GREEN) 108–9
global warming 6, 102, 103, 133–50
globalization: "conservation conundrum"
    174, 184; devolution *xvi*; economic
    development 2, 3; effect on local
    communities 92, 95; identity 121,
    124–5, 206; place-based identity 158;